*The Campus History Series*

# MILWAUKEE SCHOOL OF ENGINEERING

ON THE COVER: This image features a view of the School of Engineering of Milwaukee's Electricity Laboratory in 1918. Students can be seen hard at work gaining the "practical education" skills that school founder Oscar Werwath emphasized. At this time, the laboratory was located in the Stroh Industrial Building, which offered the perfect layout of open floor plans for the school's many laboratories. (Courtesy of the Milwaukee School of Engineering Archives.)

ON THE BACK COVER: Early-20th-century students stand outside the Stroh Industrial Building. Thousands of students have attended the Milwaukee School of Engineering since its opening in 1903. The school has a strong reputation for providing a solid technical education, which is only strengthened by its partnerships with local and national industry leaders. (Courtesy of the Milwaukee School of Engineering Archives.)

*The Campus History Series*

# MILWAUKEE SCHOOL OF ENGINEERING

Denise Gergetz and Lindsay Bastian

ARCADIA
PUBLISHING

Copyright © 2018 by Denise Gergetz and Lindsay Bastian
ISBN 978-1-4671-0354-1

Published by Arcadia Publishing
Charleston, South Carolina

Library of Congress Control Number: 2019934179

For all general information, please contact Arcadia Publishing:
Telephone 843-853-2070
Fax 843-853-0044
E-mail sales@arcadiapublishing.com
For customer service and orders:
Toll-Free 1-888-313-2665

Visit us on the Internet at www.arcadiapublishing.com

*We would like to dedicate this book to past, current, and future students of the Milwaukee School of Engineering.*

# Contents

| | | |
|---|---|---|
| Acknowledgments | | 6 |
| Introduction | | 7 |
| 1. | Electric Sparks: Early Electrical Education | 9 |
| 2. | A Low Hum: Electrical Education Grows | 27 |
| 3. | Power's On: Expanding Technical Education | 47 |
| 4. | Manmade Lightning: Educating the Public | 65 |
| 5. | Radio Waves: Radio at the School | 81 |
| 6. | Sustainable Power: Technical Education Continues | 95 |
| 7. | Building a Charge: Growth at the School | 107 |
| 8. | Current Events: MSOE Today | 121 |

# Acknowledgments

We would like to acknowledge several individuals at the Milwaukee School of Engineering who helped make this book possible. We will be forever grateful to Gary Shimek, director of the Walter Schroeder Library, for his enthusiasm and support throughout this book's creation. Special thanks are also owed to Sandy Everts, who helped get this project started. We would like to acknowledge JoEllen Burdue, Kent Peterson, and other members of the marketing department for their assistance locating images and additional historical information. We would also like to thank James Kieselburg II for his great suggestions about how to generate excitement for the book both on and off campus. We would like to express our gratitude to our library coworker Sarah Rowell who kept us smiling when the process seemed overwhelming.

We would also like to thank Jeff Ruetsche of Arcadia Publishing for initially taking an interest in our school's history, and our title manager, Angel Hisnanick, for answering our questions and keeping us on schedule.

Finally, we would like to thank our families and close friends for their patience and understanding. We have put our hearts into sharing this story of this school's innovative spirit with the world, and we appreciate all the support from those closest to us.

Unless otherwise noted, all images appear courtesy of the Milwaukee School of Engineering.

# INTRODUCTION

The Milwaukee School of Engineering (MSOE), formerly the School of Engineering, has been an important contributor to the educational ecosystem of Milwaukee for more than a century. Compelling stories about the school can be found throughout every era of its history, but the focus of this particular account is the rich series of innovations upon which it was originally founded. Although these ideas by founder Oscar Werwath were set in motion over 100 years ago, many of them have evolved with the times and many current practices are direct descendants of his original implementations.

This text focuses heavily on the earliest days of the school, from its inception in 1903 through the 1940s. It is during this time that those original, innovative ideas blossomed and helped to establish the school as a frontrunner for engineering education in Milwaukee. Those threads are then traced through the next several decades to their current iterations in the present. Some of these topics include an emphasis on a practical education curriculum, partnerships with industry leaders, education of the public, accommodating and adapting to student needs, education in wireless and radio transmission, and dedication to researching the newest technological trends.

The images, photographs, and other visual materials presented here are intended to give the reader a glimpse into the formative years of the Milwaukee School of Engineering. Numerous full-fledged histories of the school already exist, so the decision was made to narrow the focus of this text to the early innovations that made the school unique from the very beginning. This decision allowed for a more thorough exploration of our founder's originality and creativity at the turn of the 20th century. It is the authors' hope that this account sparks interest in the foundational history of the school, as well as the broader themes of the history of education in Milwaukee.

Only a narrow portion of the school's history is covered in this publication. Anyone who would like to explore the history of the institution further is invited to peruse the digitized photograph collections available through the Walter Schroeder Library's website, or make an appointment to visit the Archives of the Milwaukee School of Engineering to conduct further research.

The late 1800s saw an incredible amount of interest generated in the newest technology, electricity. Cities and transit systems were becoming electrified at an incredible rate and electricity was just beginning to find its way into households. It would not become widespread in homes for another few decades, but it was clear that this new force was here to revolutionize the modern world. It certainly caught the attention of a small German boy confined to his bed to recover from an injury.

The boy was Oscar Werwath, and some years later, he would receive a degree in electrical and mechanical engineering from Mittweida Technikum in Saxony, Germany. Reading about electrical principles and disassembling his father's pocket watches had been a childhood hobby of his, so it was only appropriate that he pursued a degree in those fields. After hearing about the booming manufacturing industries and high concentration of German immigrants in Milwaukee, Wisconsin, Werwath moved to the city in 1903. He began as an electrical engineer for the Milwaukee Appliance Company, but his true passion was teaching anyone who took an interest in the newest electrical principles. Attendance at his lectures outgrew several local venues before he purchased a building to dedicate to electrical education.

No one could have prepared for the amount of interest Oscar Werwath's school received from students, skilled workmen, and industry leaders. He handled several incredible expansions of his school in strides and continued to be the driving force behind its success for 45 years. His legacy was a school where professors took a genuine interest in their students' lives, efforts were made to educate the general public, industry leaders trusted the quality of the practical education received by students, and the most cutting-edge technological texts and equipment were made available.

Oscar Werwath's passing in 1948 left his son, Karl Werwath, as president of the school. He was extremely well-educated, having earned a degree in electrical engineering from the Milwaukee School of Engineering, and continuing his studies in business administration and education. When Karl Werwath became president, there was a period of adjustment before he found his own footing and was able to pursue his own goals for the school. The first few years of his presidency saw the creation of several research committees to record and analyze various aspects of the Milwaukee School of Engineering. Many of his father's original ideas remained in place, however, and Karl's presidency was focused especially closely on the school's industry partnerships and physical expansion of the campus.

Karl Werwath can be credited as the Milwaukee School of Engineering president who brought the school into the computer age. Like his father's certainty that electricity was the way of the future, Karl made it a point to obtain the newest computing technologies for the school and began the computer technology curriculum at the Milwaukee School of Engineering in the late 1950s. Karl shared his father's enthusiasm for educating the public about new scientific and technological developments, and had the opportunity to see that initiative through in both radio and television formats. His emphasis on continued partnerships with industry, campus expansion, and new technologies set the stage for the school to thrive into the modern era.

The Milwaukee School of Engineering blossomed under more than 75 years of leadership by the Werwath family. The tradition of strong leadership has continued through the presidencies of Dr. Robert Spitzer, Dr. Hermann Viets, and Dr. John Walz. Each of these men have helped the school become a nationally recognized institution known for graduating students with a unique blend of practical and theoretical education. For more than a century, the Milwaukee School of Engineering has held a reputation of excellence. The solid foundation of innovation set forth by founder Oscar Werwath and the continued leadership of innovative minds allows the school to look to the future with confidence.

## *One*

# ELECTRIC SPARKS
## EARLY ELECTRICAL EDUCATION

In 1889, at the age of nine, Oscar Werwath was thrown from the horse he was riding, which left him with a spinal cord injury. He was bedridden for several months as he recovered from the accident. Unable to do much else, young Oscar read anything and everything he could find about the emerging field of electricity. When he was not reading, he entertained himself by taking apart and reassembling his father's pocket watches. Oscar did not know it at the time, but these budding interests would shape his life and echo into the next century.

Werwath received his engineering degree at Mittweida Technikum in Saxony, Germany, in 1898 and immigrated to the United States only a few years later. He chose to settle in Milwaukee because of its reputation as a center of skilled labor and high-end industry, and was happy to become a part of its large population of German immigrants. It was during his time working as chief designer and consulting engineer with the Mechanical Appliance Company of Milwaukee (Louis Allis Company) that he first began holding informal lectures around his kitchen table for fellow engineers, foremen, and workers. Werwath would share his knowledge about the emerging field of electricity and teach interested colleagues how to build crystal radio sets. Attendance at these lectures soon outgrew his kitchen.

For two years, Werwath lectured at Rheude's Business School, using his own notes from college. When attendance at his lectures outgrew this space as well, he purchased a former store building at 1025–1027 Winnebago Street in Milwaukee and opened the School of Engineering. The focus of his school was on practical education—applying principles for creative problem solving, engaging in hands-on learning, and emphasizing the training needed to obtain skilled employment. This proved a very successful approach to education, and by 1905, a total of 100 students were enrolled at the school. The following year, that number more than doubled.

Local manufacturers took notice of the School of Engineering as well. Industry leaders were quick to realize that employees who worked part-time and attended the school part-time gained crucial skills much faster than those taking part in seven-year apprenticeships. As the backgrounds of the men enrolling grew more and more diverse, Werwath was quick to add a machine shop and electrical, chemical, and physics laboratories to the Winnebago Street building.

Pictured here is the first home of the School of Engineering on Winnebago Street in Milwaukee. Originally, the upper floor contained living space for Oscar Werwath and his family and classrooms for evening lectures. This building was renovated in 1906 to include a machine shop in the basement where students were encouraged to manufacture devices and equipment of their own design.

The first students to attend classes at the School of Engineering were young men looking to advance in technical jobs. The first instructors at the school were predominantly foremen and workmen at the shop of Julius Andre & Sons where Werwath had worked when he first immigrated to Milwaukee. Pictured here are early instructors and students lined up outside the Winnebago Street building in 1907.

School of Engineering founder and president Oscar Werwath poses outside the Winnebago Street building in his all-electric vehicle. His wife, Johanna (Seelhorst) Werwath, noted later in a scrapbook that this car was nicknamed "the Pride of 1905." By 1918, the school would acquire two more buildings, one of which housed the expanding School of Automotive Electricity.

Oscar Werwath was extremely skilled at promoting his school both to local industries and prospective students. His enthusiasm for the subject matter and ability to teach it to anyone made him a successful advocate of the school. Here, two students pose with a car bearing a "Milwaukee SOE" pennant and a sign proclaiming the School of Engineering of Milwaukee the "largest electrical school in the world."

The faculty members at the birth of the School of Engineering were serious about education, but that did not mean there wasn't the occasional opportunity for fun. Oscar Werwath can be seen seated at center. Johanna, his wife, later noted that this was a well-deserved evening of levity after weeks of hard work in 1905.

Oscar Werwath enjoyed hosting events for both faculty and students. Here, he can be seen at the rear of the table at a 1905 dinner for faculty members in his bachelor quarters at the Winnebago Street building. The environment of camaraderie he fostered at the School of Engineering is recalled with fondness in the writings of students and faculty alike.

In the Mechanical Engineering Laboratory, two students can be seen adhering to Oscar Werwath's philosophy of hands-on learning. After completing an experiment, a student would write up his findings. If his conclusions were incorrect, he was given another chance at the report. The intention was that the student would amass a series of laboratory reports correctly discussing important electrical principles, which he could refer to after graduation.

In 1905, Milwaukee was known for its large population of German immigrants and was a bustling center of manufacturing. It was the perfect place for Oscar Werwath to set up his school, as his unique vision for providing a practical education offered a direct route into skilled employment. Pictured here is the Electrical Testing Laboratory, with wiring experiments lining the walls and electric motors in the center.

The Chemical Laboratory was considered a part of the Institute of Electrotechnics, and the work undertaken there was applied directly to engineering education. Chemistry students would progress from classes in general chemistry to industrial chemistry, to electrochemistry in the College of Electrical Engineering. This image shows the first chemical laboratory at the School of Engineering.

A promotional booklet for the school, *Photo Story*, describes the Chemical Laboratory as a bustling place full of students conducting experiments. The flashes of light and sounds of generating and burning oxygen, exploding hydrogen gas, and analyzing alloys are described in breathless excitement by a fictional prospective School of Engineering student.

This machine shop was in the basement of the Winnebago Street building. Students gained practical experience manufacturing electric machines, meters, storage batteries, switchboards, and other electric appliances here. The revenue from selling these items was used to offset tuition costs and build up the school's electrical laboratories. This was the first of many efforts by Oscar Werwath to help students with rising tuition costs.

Students are working in the Electrical Laboratory at the Winnebago Street building in 1910. The two large pieces of machinery on the left table are generators designed and built by School of Engineering students. They were used to power wiring laboratory work, which was conducted on the board on the back wall.

Two students are pictured repairing storage batteries. Building and repairing storage batteries—especially those used in automobiles—was a way that the School of Engineering raised money in the early years. This battery business would eventually be absorbed into Globe Battery, and a strong partnership grew between the school and the company.

As the electricity and storage battery industries grew at the start of the 20th century, businesses needed men trained in these technical fields. The School of Engineering bridged this gap by teaching students the emerging skills employers were seeking. Here, four students are learning about storage batteries. Such practical knowledge would be useful at an automobile service station or automotive garage.

Pictured here is a large storage battery of the type that would have been used on a farm or small-town power plant around 1910. It was difficult to get power out to the rural areas of Wisconsin, and often smaller entities needed to build or purchase a power plant of their own. The School of Engineering produced and sold these batteries as well as lighting setups and power plants.

A similar storage battery was a crucial part of this electric lighting plant, which used the newest patented low-voltage tungsten lamps available at the time. The School of Engineering sold these plants and provided hundreds of installations to homeowners in rural areas, which gave students the opportunity to gain experience by taking part in the installations.

In 1907, a School of Engineering student could expect to spend half his time in lectures learning theory, and the other half in the laboratories putting those theories into practice. The Winnebago Street building contained a commercial parts store where the public could purchase the batteries and items assembled by students (above) to offset equipment costs. The school also offered a store where students could purchase equipment and parts for their own laboratory experiments. In the image below, this store has been decorated for the 1906 school exhibition that was open to friends, family, and the public.

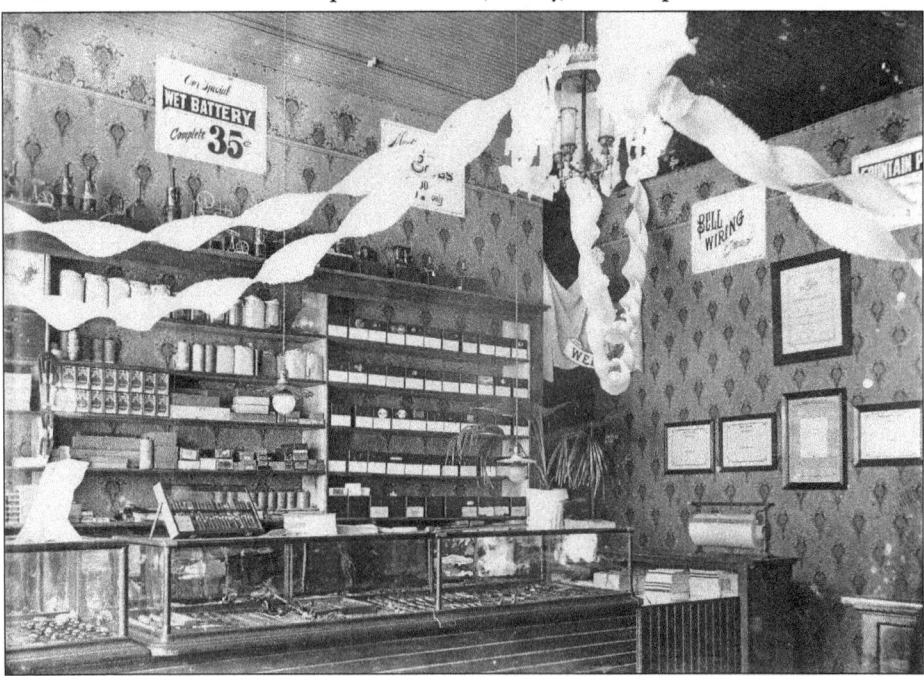

Oscar Werwath made every effort to connect with the public, and one annual tradition was the Great Exhibition at the School of Engineering. At this open house, students showcased the products of their classwork throughout the Winnebago Street building. In this image from 1906, many of the dynamos and motors in the Electrical Laboratory are on display beneath a chalked sign intended for the less tech-savvy visitors.

Pictured here is the Drawing Room decorated for the 1906 Great Exhibition. The oversized slide rule displayed beneath the chalkboard is identical to those that hung in engineering classrooms until recently. The label above it has been placed for the benefit of the public. Students were proud to show off their drawings; often, these designs were used by students to build original machinery and apparatuses in the school shops.

Armature winding for electric motors and generators was a subject that received much attention throughout the training of students. Every student was required to wind and test several different types of armatures from start to finish. This included winding the coils on a loop winder, forming the coils on a coil spreader, insulating the coils with tape, placing the coils into the armature slots, soldering the leads to the commutator, testing the assemblage for short circuits and grounds, and (if it was a direct current motor) to finally turn down the commutator in the lathe. Before a student could progress to the next step, the instructor would evaluate his work to ensure the proper care and attention had been given to the task.

Practical knowledge about emerging technical fields was highly sought after in the early years of the 20th century, and it did not take long for student enrollment to increase beyond the capacity of the Winnebago Street building. In late 1911, Werwath moved part of the school to the Stroh Industrial Building at 161–171 Michigan Street. The 1912 course listing calls the building "Milwaukee's newest, brightest, and most modern concrete fire-proof building."

Pictured here is one of the bright new drafting rooms in the Stroh Industrial Building around 1915. Students enrolled in this mechanical drawing course could expect to learn the application of mechanical drawing to electrical apparatus, as well as the practical skills of drawing freehand lettering, geometrical problems, and sections and intersections.

The Stroh Industrial Building contained a fully stocked Alternating Current Testing Laboratory. The experiments students were to perform here were initially very simple and grew more complex as students gained confidence and experience. One early experiment was designed to teach the student how to make simple connections of bells, push buttons, switches, and electric lights. These tests were intentionally constructed in a manner similar to that in which homes were wired.

The School of Engineering's first academic library appeared in 1912 in the Stroh Industrial Building. Previously, there had been an informal book exchange in the student lounge at the Winnebago Street building, but it was intended for popular reading materials and not scientific texts. The Stroh library was the first school facility dedicated to periodicals and textbooks related to the subjects studied at the school.

In this image from 1912, Prof. Fredrick C. Raeth stands in the Electro-Chemical Laboratory in the Stroh Industrial Building. In the school bulletin from 1912, Professor Raeth was heralded as one of the most capable physical directors in the city because of a string of successful sports team victories under his direction.

This 1912 lecture on high frequency was accompanied by a demonstration of the wireless monoplane pictured at right against the windows. The airplane model was built by alumnus Edward Koeppel, and was completely controlled via a wireless control key. In the summer of 1913, this lecture and demonstration would leave quite the impression upon attendees at the Milwaukee Electrical Show, managed by Oscar Werwath.

Oscar Werwath worked hard to foster an environment of friendliness at his school and made an effort to know each of his students on a first name basis. This was possible because he taught a full schedule of classes up until 1918. Managing the growing school began to take its toll on his workload, and he had to focus his attention on administrative duties full-time. He did, however, continue to teach a single foundation course in general electricity until 1928. In that way, he was able to remain a part of students' educational lives. Above, two students work with electric motors in 1910. Below, two students from 1910 test electrical circuits.

# CHAPTER V.

## INSTALLATION OF BELLS AND ANNUNCIATORS.

*The Tool Kit.* — In order to properly install any electrical apparatus, it is necessary to have the proper tools with which to perform the work, and in buying these, it is the cheapest in the long run to get the best. It is also essential that they be kept in the best of condition. The saws and chisels and such cutting-edge implements should be kept sharp and free from rust. See that all necessary tools and supplies, such as screws, wire and tape are in the tool bag before starting out on a job, as this will effect a great saving of time.

After careful consideration, the following list of tools and supplies is given, as fulfilling the needs of the electrician who intends to do work installing bells, and so forth.

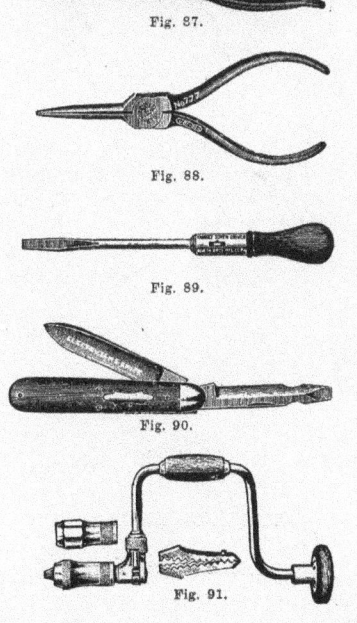

One eight-inch side-cutting pliers.

Fig. 87.

One long-nose pliers.

Fig. 88.

One screw driver with three-inch blade.
One screw driver with six-inch blade.

Fig. 89.

One pocket knife.

Fig. 90.

One brace.

Fig. 91.

---

Oscar Werwath established the Electroforce Publishing Company on the first floor of the Stroh Industrial Building in 1912. He worked closely with the Krus Engraving Company to design and print the complex photographs and drawings necessary for his vision. Werwath's first major publication was the technology-focused journal *Electroforce,* which he marketed as "the Technical Magazine for Everybody." Over the next several years, and with the help of faculty members, his endeavor grew into a series of 16 textbooks about practical electrical education. This is a page from his first volume in 1919, *School of Practical Electricity Book I: Fundamentals of Electricity and Wiring*. Werwath's works were quickly put into use by other technically focused colleges, and even some high schools, because of his use of clear illustrations and easy-to-follow explanations.

Laboratory panels were in use at the School of Engineering from the very beginning. It was easy to rearrange laboratory areas in the cramped Winnebago Street building when experiments and equipment were mounted on portable boards that could be wired to each other in various ways. After Oscar Werwath established his publishing company, it was only a matter of time before he began to market the laboratory panels, textbooks, and equipment together for students wishing to learn practical electricity through the mail. The Milwaukee School of Engineering Correspondence School was not the first of its kind, but it was the only one to offer a complete home laboratory for students at a distance. It was advertised in publications of the day as the only series of correspondence courses offering "real" equipment to be mounted on boards for home experiments, as opposed to scaled-down versions offered by competing schools.

# *Two*

# A Low Hum
## Electrical Education Grows

The latter part of the 1910s and into the 1920s was a period of incredible growth at Oscar Werwath's engineering school. Efforts by President Werwath to market his school outside southeastern Wisconsin had come to fruition, causing him to add the phrase "of Milwaukee" to its name.

World War I had an effect on the school as well. There was the need to create a branch of the Students' Army Training Corps at the school, which required the purchase of a new building to provide barracks for the men. The four-story Marshall Street building was acquired and quickly remodeled to make it ready for habitation. When the war finally ended, the influx of young men returning home increased the school's need for instructors along with laboratory and classroom space. The Marshall Street building was then remodeled to house the Schools of Practical Electricity and Automotive Electricity.

Finding space for all the new enrollees was not the only challenge Werwath faced at this time. Classroom instructors numbered 13 in 1918 and approximately 33 in 1919; in preparation for the 1920 school year, Werwath added an incredible 76 new teachers. The school was divided into four distinct disciplines, each of which needed instructors, classroom and laboratory space, and specialized equipment.

One of Werwath's defining ideas would come out of the challenge of finding space for everyone. The "Earn While You Learn" program was his answer—students attended classes one week and gained experience working in industry the next. Another section of students was on the opposite schedule. This innovation would lay the groundwork for future partnerships with industry while offering students the chance to gain hands-on experience and have a source of income to pay for tuition.

A predecessor to the Concentric Curriculum appeared soon after, allowing electrical engineering students to participate in one of three course tracks. Basic electrical classes were taken by all, but after the first classes were completed, a student could choose to continue his education or enter the workforce. Werwath was constantly searching for ways to accommodate different levels of learners, and he worked hard to address the needs of each individual student.

In an electrical laboratory in the late 1910s, students use unit test boards to determine measurements of resistance in series and in parallel. These unit boards used the same equipment as those sold as part of the correspondence course that Werwath had established a few years earlier. Distance students simply had to assemble the pieces and provide their own mounting board.

These students in 1920 are testing circuits in the recently acquired floors in the Insurance "Loyalty" Building on the corner of Broadway and Michigan Streets. At this time, this building served as mostly administrative and departmental offices, but it also contained a few electrical engineering laboratories.

Although the school was changing quickly, Oscar Werwath's goal of providing young men with a practical and technical education remained consistent. Both of these images show experiments in which very feeble electrical currents are measured by galvanometers, and resistance is measured via a Wheatstone bridge. In the photograph above from 1907, a mirror galvanometer, a tangent galvanometer, and a Wheatstone bridge are pictured in an experiment space in the Winnebago Street building. The photograph below, taken in the 1920s, shows two students gathering similar data using a slightly more modern version of the Wheatstone bridge and a galvanometer.

The School of Engineering of Milwaukee formally opened its newest building on March 17, 1917. The school inhabited three floors of the Insurance "Loyalty" Building, located at 373 Broadway. The entire engineering department moved to this location, which gave the electrical department room to expand its laboratories in the Stroh Industrial Building. In this image of the building from 1919, the Students' Army Training Corps marches in formation in the street below.

In 1917, students of the newly chartered College of Electrical Engineering pose on the impressive main staircase of the Insurance "Loyalty" Building. They would be the first group of School of Engineering of Milwaukee students to have the option of obtaining a bachelor of science in electrical engineering.

Senior students enrolled in the day classes offered at the School of Engineering Institute of Electro-Technics pose proudly outside the new Insurance "Loyalty" Building in 1917. Although the address has changed since this photograph was taken, this beautiful building and its arched doorways can still be seen in Milwaukee today.

In 1917, students could focus their studies in any one of four areas. The College of Electrical Engineering was available to high school graduates and offered a 12-term course in electrical engineering. The Institute of Electro-Technics was for non–high school graduates and offered an eight-term course. The School of Practical Electricity offered a complete course in six months, and the School of Automotive Electricity offered a complete intensive course in three months.

This photograph from 1918 shows Oscar Werwath's office in the Insurance "Loyalty" Building. The rapid rate at which the school was expanding was a credit to his vision, energy, and aptitude to create a school recognized as one of the most practical electrical institutions in the United States. The students he had educated were taking part in the growing electrical industry all over the world.

This image from 1918 represents the last time the teaching staff at the School of Engineering of Milwaukee would be this small. After World War I, the influx of young men returning home to look for work would cause enrollment at the school to grow exponentially. School founder Oscar Werwath is seated third from the left.

Esther Shapiro was the School of Engineering of Milwaukee's first female teacher. Shapiro graduated from East Division High School, where she specialized in mathematics, in 1911, and from the University of Wisconsin in 1915 with a bachelor's degree and a teaching certificate in mathematics. According to her surviving relatives, her father dared her to apply for the position at the School of Engineering, as he knew that the school had only male teachers, but that many of them had gone off to war. Shapiro accepted his challenge and was hired in 1918 to teach mathematics. It was noted in the school paper, *Electroforce*, that the hiring of a female professor was a welcome and progressive change for the school. Esther Shapiro was well-liked, as evidenced by several affectionate student-drawn comics published in the campus yearbooks and the school newspaper. This illustration, entitled "Cross-Section of Engineering II Math Class" originally appeared in the 1919 student yearbook.

Pictured here is the School of Engineering of Milwaukee faculty for the 1919–1920 school year. When the United States entered World War I in 1917, many schools and universities became training centers for Army officers. This meant that curriculum material was decided upon and highly regulated by the government, which kept faculty numbers stagnant through 1918. A section of the Students' Army Training Corps was set up at the school in October 1918 and housed in a four-story building acquired for that purpose. Once the war ended in late 1918, the school received government recognition and was permitted to organize a unit of the Reserve Officers' Training Corps. This event is credited with a significant uptick in enrollment, with the school requiring approximately 20 new faculty members for the 1919 year and a staggering 76 new faculty members for the 1920 school year.

In addition to collaborating with the government to serve its needs for technological education, the School of Engineering of Milwaukee has always cultivated relationships with local business leaders. This advertisement from the back of a 1913 course catalog highlights a cutting-edge addition to the school's electrical laboratory—a complete telephone exchange—donated by the Wisconsin Telephone Company. Oscar Werwath worked closely with prominent leaders in Milwaukee industry to build technical programs specific to their needs. Once these leaders saw the quality of the skilled graduates produced, they were often willing to donate specialized equipment to prepare the men for jobs in their unique industry. This was a mutually beneficial alliance, as the School of Engineering received the newest technologies with which to teach students, and the industries received an unending supply of already trained graduates. Partnerships with specialized industries, like this one from the early years of the school, were an innovative effort at the time, and the importance of amiable relationships with industry leaders is still a significant focus of the school today.

The growing electrical laboratory also received a seven-pole transmission line, complete with all transformers and electrical equipment, as well as electrical meters and testing equipment from the Milwaukee Electric Railway and Light Company in 1913. Electrical power for the transmission line and the rest of the laboratories was provided by a 2,300-volt generator donated by a former student.

This image shows students around 1920 conducting laboratory work in the meter testing and calibrating department. The ability to use and calibrate delicate instruments for measuring electrical currents was a necessary skill for every graduate, no matter which part of the electrical industry he would enter after graduation.

The four-story Marshall Street building was purchased by the School of Engineering of Milwaukee in 1918 to house the newly formed Students' Army Training Corps. The building underwent a complete renovation in only a few months. The first floor held offices, kitchens, dining areas, and a library. The second floor served as sleeping quarters and had an isolated infirmary area. The third floor provided living space, washrooms, and showers.

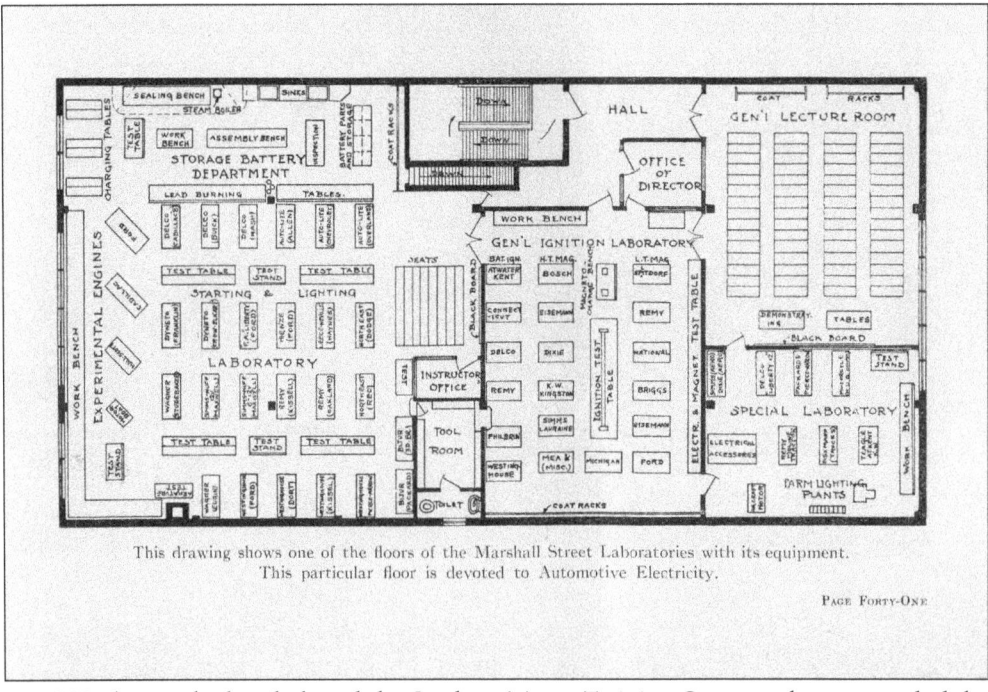

This drawing shows one of the floors of the Marshall Street Laboratories with its equipment. This particular floor is devoted to Automotive Electricity.

By 1920, the war had ended, and the Students' Army Training Corps no longer needed the space for housing. The School of Engineering of Milwaukee, with enrollment still increasing, repurposed the building yet again to house the School of Automotive Electrotechnics. This building plan depicts the main starting, lighting, and ignition laboratories.

In 1927, two students demonstrate the power of an electromagnet. Using only two dry cell batteries, it was said to be capable of lifting 1,500 pounds. Generally, students would determine the electromagnet's lifting capability by means of a spring balance attached to the armature.

Here, five students of practical electricity conduct laboratory tests on alternating current machines in the 1920s. Oscar Werwath's dedication to marketing and outreach certainly paid off after the war. During this time of tremendous growth for the school, the student body boasted members from many states and more than 20 different countries, including China, Korea, and Australia.

Pictured here in 1920, the Marshall Street building housed the School of Automotive Electricity. The location was perfect for this purpose because it was a mere two blocks from Milwaukee's "Motor Row," an area of the city known for automobile agencies, service stations, and repair shops.

Students of the School of Automotive Electricity can be seen working in the General Ignition Laboratory in 1920. This laboratory was said to contain mounted versions of every type of ignition system found on gas-driven vehicles of the time. Students were assured that they would finish their courses with the ability to maintain and repair any type of magneto- or battery-ignition equipment.

Oscar Werwath's innovative focus on practical education extended to every department at the School of Engineering of Milwaukee. He made sure that a student's time spent applying lecture theory was done in conditions that mimicked real-world situations. In this image from 1920, practical electricity students practice wiring a model house according to National Board of Fire Underwriters regulations.

In addition to automotive and practical electricity laboratories, the Marshall Street building also had two drafting rooms. This image from 1920 shows students hard at work in the smaller of the two drafting rooms. These students would have been high school graduates just starting their course in electrical and mechanical drawing in the Institute of Electrotechnics.

Although practical and technical training were heavily emphasized at the School of Engineering, Oscar Werwath also recognized the value and importance of research. So, in addition to completing the practical curriculum and passing their examinations, some students, particularly those completing higher-level degrees, were asked to submit a thesis. Students worked with an assigned faculty member to determine an area of study and completed this report over the course of their final year at the school. These reports often focused on ways that electricity could be used to improve existing technologies. Topics included electrifying railways, street-lighting systems, improving public utilities, and motor patents. As an incentive to do thorough work, a gold medal was awarded to the student with the best thesis. Some of the surviving reports have been preserved in the Archives of the Milwaukee School of Engineering.

Life at the School of Engineering of Milwaukee was not all classwork. After the war, two semesters of physical education instruction became mandatory for all students. The 1920 *Photo Story* promotional booklet states, "Our primary purpose is to secure the best physical training for all our boys. By our method spindle legs, sunken chests, and shambling walks are transformed into strong muscles, proper carriage, and manly bearing."

Promotional materials for the school in the early 1920s began to highlight efforts to improve the whole student in both mind and body. Considered a "clean manly sport" by the administration, basketball was especially promoted to the student body as it could be played with equal ease at any time of the year, inside or outside. Other sports that promoted honorable play and healthful recreation were baseball, football, and boxing.

In 1917, the School of Engineering of Milwaukee had four institutes a student could choose from based on his situation—from the three-month program in the School of Automotive Electricity to the three-year course of study in the College of Electrical Engineering. By 1926, these different courses of electrical training were advertised to prospective enrollees in pamphlets, simplified in the shape of a pyramid. Students wishing to quickly gain the skills needed for technical employment could opt out of the full two-year course and still know that their education in electricity was complete. The two-year timeline was the ideal situation and applied to those students who had graduated high school and could dedicate full-time hours to their studies. This pyramid diagram lays the groundwork for the later concept of the Concentric Curriculum, an innovative formal diagram of the flexibility in School of Engineering course paths created by Oscar Werwath.

During this time of growth, Oscar Werwath also used his industry connections to implement the Earn While You Learn program. It grew out of the student assembly and sale of batteries during the Winnebago Street years. This time, however, students could sign up to work part-time for one of the school's industry partners. Often, this would occur on a week-on, week-off basis—one student would spend a week working and the next in lectures. Then, while he was attending lectures, his shift at the company would be taken by a fellow student who would attend lectures the opposite week. The image above shows School of Engineering students gaining practical experience in the Globe Battery workshop in 1924. Below, a student in the late 1920s takes part in the program by repairing radio sets in a local shop.

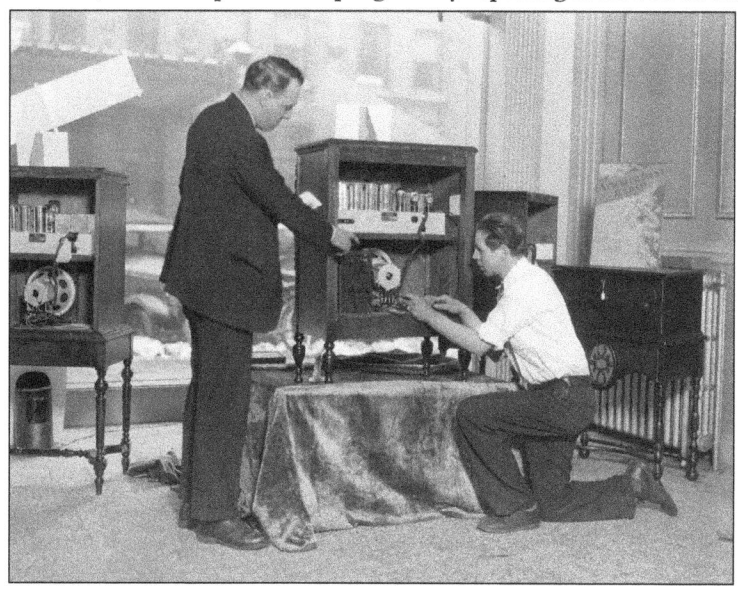

Although by the late 1920s the population of students and faculty at the School of Engineering of Milwaukee had grown far too large for Oscar Werwath to entertain in his home, he still strived to keep the friendly small-school atmosphere alive. Here, members of the school's various athletic teams pose together after a school-sponsored formal "mixer" event.

As the school celebrated its 25th anniversary in 1928, the rapid physical expansion brought on by World War I began to take on a less-urgent pace. This image from the Silver Anniversary celebration on November 23, 1928, shows students seated in the middle divided by fraternity or club affiliation, and faculty members at the head table against the wall beneath the American flag.

# *Three*

# POWER'S ON
## EXPANDING TECHNICAL EDUCATION

The rapid physical expansion of the 1920s had ended, and Oscar Werwath's School of Engineering of Milwaukee had established its place on a national scale as a solid institution for a young man to receive an electrical education. However, the administrative duties had become more than Werwath could handle on his own, and he formed the Industrial Advisory Board and the Board of Regents, and established the school as a nonprofit organization.

Enrollment slowed as the effects of the Great Depression set in, and the school began to look for new ways to diversify its course offerings. Classes that would eventually grow into the Refrigeration, Heating and Air Conditioning Institute were added to the curriculum in 1926, and in 1933, Werwath added the Welding Institute. This proved a smart addition, as World War II would increase demand for men with this specialized skill. Both the three- and six-month welding programs were extremely popular, and during the war, evening classes in welding were made available to women.

Aviation instruction was also added to the curriculum during the war. Although the school could not grant pilot certifications, aerospace topics bled into regular coursework. A source of pride for Werwath was that a type of glider built at the school had been purchased by the military and had landed in enemy territory.

After the war, the Refrigeration, Heating and Air Conditioning Institute was added to the school, and a formal class track of radio and television instruction was established. Radio was always a pet project of Werwath's, and it must have been gratifying to see it blossom into a formal set of classes.

Oscar Werwath passed away in 1948. He had watched his school grow from offering a few classes to teach working technicians new skills, to a nationally recognized and degree-granting institution. His success is a credit to his dedication to the idea of "practical education," his fascination with new technologies, the partnerships he forged with industry leaders, and above all, his devotion to each and every student who passed through his school.

The Oneida Street building (originally located on the south side of what is now Cathedral Square Park) was acquired in 1926 and used to house the school's radio broadcasting equipment and studio, as well as a handful of offices and classrooms. Despite the acquisition, physical expansion of the School of Engineering of Milwaukee slowed as the effects of the Great Depression began to affect students and faculty alike.

During efforts to establish a student loan fund in 1932, the Industrial Advisory Board formed the Corporation of the Milwaukee School of Engineering. The group then formalized the principles of practical education on which Oscar Werwath had built the school by signing the Articles of Incorporation. To celebrate the school's place as a nonprofit and established institution, the Board of Regents purchased the German-English Academy building for the campus. This photograph is from 1979.

In conjunction with the signing of the Articles of Incorporation in 1932, the name of the school was officially changed to the Milwaukee School of Engineering. While this was certainly a proud moment for founder Oscar Werwath, the dire financial climate of the decade would force him and his advisors to find innovative new ways to appeal to potential students.

Pictured here in the 1930s, Oscar Werwath (fifth from left) and his industrial advisory committee stand outside the School of Engineering of Milwaukee. At this time, the school had grown too large for Werwath to manage alone. Out of this challenge, he created several advisory boards and the Board of Regents and was approved to run the school as a nonprofit.

# CONCENTRIC CURRICULUM

The Concentric Curriculum was one of Oscar Werwath's innovations from the early days of the school, and the concept was first referenced in 1918. Werwath conducted several investigations into the expectations of the student body and found that the practical education that students wanted was radically different from those classes offered by other educational institutions. Initially, the Concentric Curriculum was a way for students with limited time or money for schooling to find a technical program that fit their career goals. In the mid-1920s, the concept materialized in promotional materials as a pyramid—every student received the foundational classes and could continue his education according to his needs. Finally, the idea was illustrated in school bulletins as a series of concentric circles, showing prospective students the courses of study required for practical education, a job as an electro-technician, or an electrical engineer. The example reproduced here comes from the 1947–1948 Annual Catalog and includes additional classes that would be added to the curriculum in the coming years.

The goal of the Concentric Curriculum was simple: to consider the aptitude of each individual student and his ability to absorb the class material. His abilities and aptitude would determine which course path would best fit his needs. Pictured here are students determining the effect of condensive action in alternating current circuits. Students of all levels began this type of practical training early in their coursework.

Flexible options were available for completing programs at the Milwaukee School of Engineering. In one year, a student could complete the Commercial Engineering unit; in two, he would finish the Industrial Electrical Engineering unit; and in three, he would graduate with a bachelor of science degree in electrical engineering. These electrical engineering students in the early 1940s are learning how to read blueprints.

Both of these photographs are set in the same open laboratory space in the German-English Academy (Broadway) building in 1942. The airy, open floor plans of many of the Milwaukee School of Engineering laboratories allowed for cross-pollination of scholarly disciplines, materials, and equipment. These innovative setups are a physical embodiment of Oscar Werwath's Concentric Curriculum idea, as the commercial and industrial engineering paths shared many classes with the longer bachelor of science program. This not only allowed the students to share resources, but to also gain experience interacting with differently trained men within their chosen field. Werwath's emphasis on not only scholarly education and practical experience, but also professional relationships, was one of the most influential factors in making his school unique.

Not only was the laboratory space at the Milwaukee School of Engineering intended to encourage students to collaborate, but there was some significant crossover with the equipment necessary to complete laboratory assignments. Early on, this was remedied by attaching large motors and dynamos to transportable bases so that they could be moved to whichever laboratory space required them. As this type of technology grew smaller, the school was able to store multiple examples of the same piece of equipment in special rooms for students to check out as needed. The image above shows a student replacing a fractional horsepower motor on the shelf after an experiment in 1947, and below is a view of another equipment room filled with meters and testing apparatus.

In the 1930s, the popularity of evening classes at the Milwaukee School of Engineering grew significantly. Many of these courses were only three or six months in length and intended to teach practical skills, like this lecture from instructor Donald Downing about welding. Many of the students enrolled in these classes were employed in Milwaukee industries and looking to learn new skills.

Oscar Werwath added the Welding Institute to the Milwaukee Schools of Engineering in 1933, and by 1948, it boasted having trained 5,500 welders and welding technicians. This promotional bulletin from 1938 was distributed by the school to interest potential students in gaining new skills. Although welding advanced at the start of the 20th century, World War II pushed welders to innovate new techniques quickly.

Safety was the prime consideration in the Welding Institute. In this photograph from 1940, students dressed in safety equipment line up to collect their supplies for the day before entering the workshop. Although the practice of welding has changed since the 1940s, the need for protective leather aprons, gloves, and helmets remains the same today.

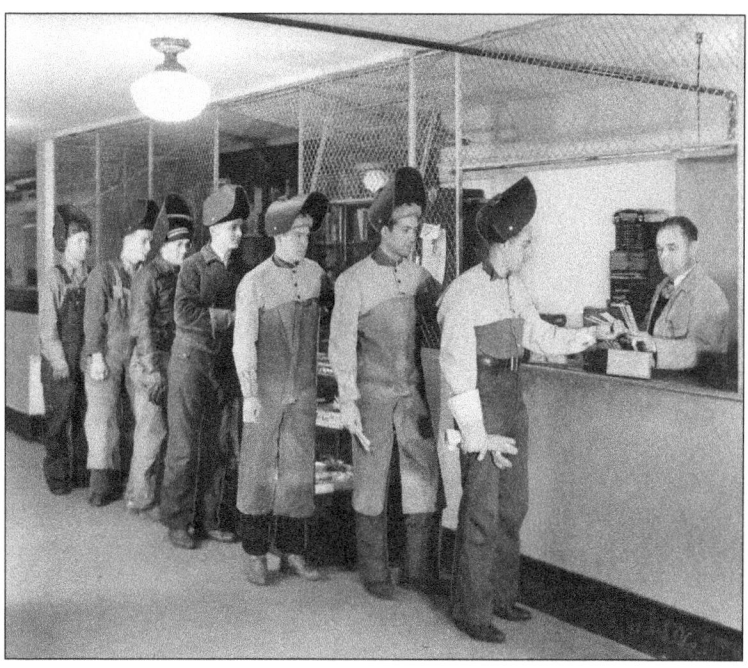

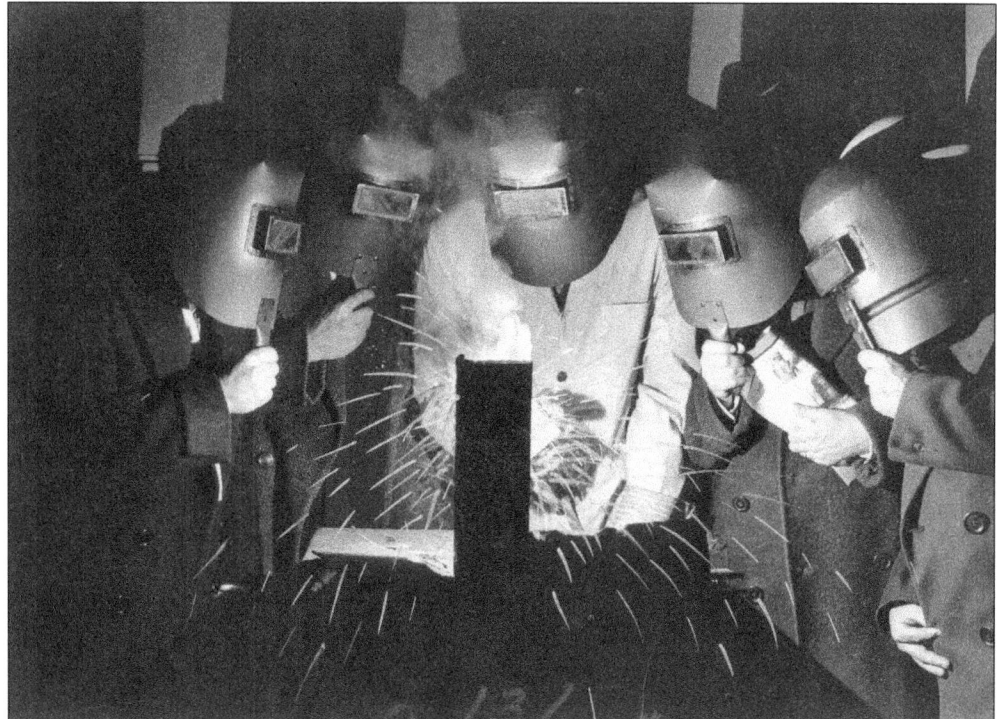

In this photograph from the early days of the Welding Institute, a student demonstrates the art of vertical welding to a group of onlookers. The demonstration was credited in the school bulletin as one of the most interesting at the school's public open house that year.

The onset of World War II caused enrollment to decline at the Milwaukee School of Engineering as many young men went off to fight in the war. Taking a cue from his success updating the curriculum to better serve the needs of skilled workers during World War I, Oscar Werwath began to shift the school's focus to address skills needed in the military. Although the school could not offer flight training to pilots, Werwath adjusted many of the courses to address the needs of the aviation industry. This image from 1940 shows two young men who are a part of the school's three-month concentrated course in aircraft welding. A two-year aerotechnician program was also added, which focused on wood fabrication, aerodynamics, navigation, meteorology, and aircraft assembly, rigging, and instrumentation. These classes ran until 1944.

During World War II, the laboratories of the Milwaukee School of Engineering were filled with cutaway models of planes, motors, wings, propellers, and aircraft instruments. This equipment allowed for the complete study of aircraft and airplane construction. Under the watchful eye of their professor, these students study the construction of an airplane using a cutaway model.

Staying true to his dedication to teaching members of the public, Oscar Werwath continued to hold school open houses during World War II. It is no surprise that the plethora of cutting-edge aviation equipment in the laboratories captivated visitors. Seen here are several members of the public looking on as an instructor speaks about the components of an aircraft frame in 1940.

No history of innovation at the Milwaukee School of Engineering would be complete without mentioning the glider project. A particular source of pride for Oscar Werwath, a glider was designed by participants in the aerotechnician program at the school and subsequently purchased for use by the US Army Air Force. The school built several of the gliders, and sometime later received word that the machines had, indeed, landed in enemy territory. The photographs here, both taken in 1941, show aspiring aerotechnicians assembling a few of the gliders. Werwath can be seen admiring their work in the background at right below.

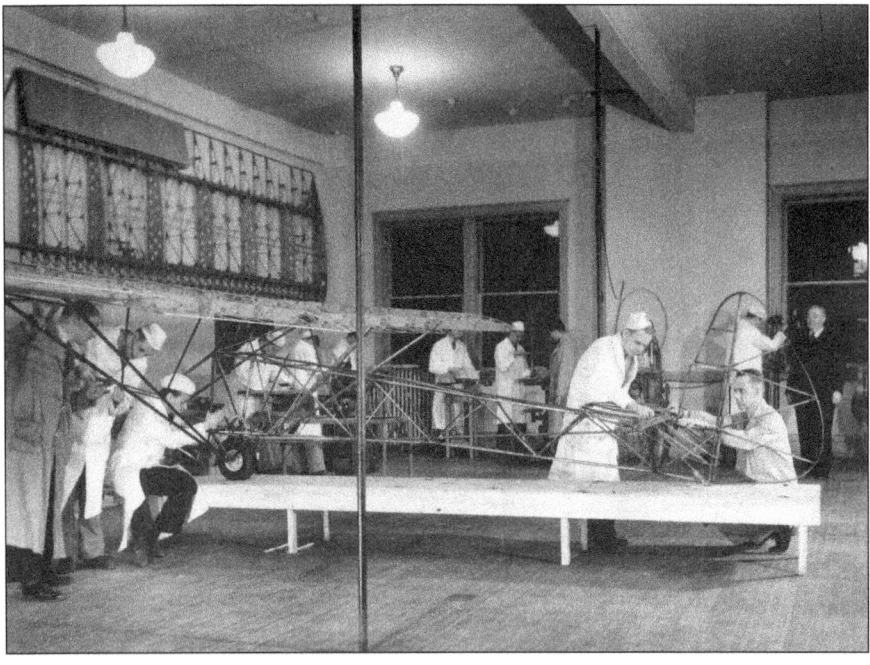

By 1942, the Milwaukee School of Engineering's courses were solidly tailored to the needs of the military. Even the courses that had been a part of the school since the earliest years—electricity, radio, drafting—received a makeover to better align the competencies they taught with skills that would help in the war. On this promotional booklet from 1942, there is a definite appeal to those young men eager to do their part in the war effort, and an emphasis on jobs related to airplane building and maintenance, drafting, and communications. Inside the booklet, Oscar Werwath not only assures potential enrollees that skilled technicians will quickly ascend the military ranks, but he also frames the classes as future-focused. He acknowledges that technology advances quickly during wartime and appeals to the need for new technical skills once the world returns to peace.

Although introductory classes were added to the Milwaukee School of Engineering course offerings as early as 1926, it was after the war that the refrigeration, heating, and air conditioning industry absolutely took off. Oscar Werwath was quick to determine that the school needed to dedicate more resources to this emerging field. By 1945, he had added a two-year program and the Refrigeration, Ventilating, Heating, and Air Conditioning Institute.

Pictured here in the mid-1940s, students of the Refrigeration, Ventilating, Heating, and Air Conditioning Institute work in a laboratory. Graduates could expect to find careers in contracting and service, system design, manufacturing, and equipment sales. One of Oscar Werwath's true skills was predicting which industries and technologies would make the biggest impacts on society. He worked tirelessly to train skilled men for the new jobs.

The Refrigeration, Ventilating, Heating, and Air Conditioning Institute was so successful for the Milwaukee School of Engineering that promotional booklets were created focusing specifically on that institute. This photograph, featured in the 1948–1949 *Photo Story*, depicts two students making a temperature determination on a refrigerator cabinet. They are calculating the length of time it takes to reduce the chest-type home freezer's temperature to storage level.

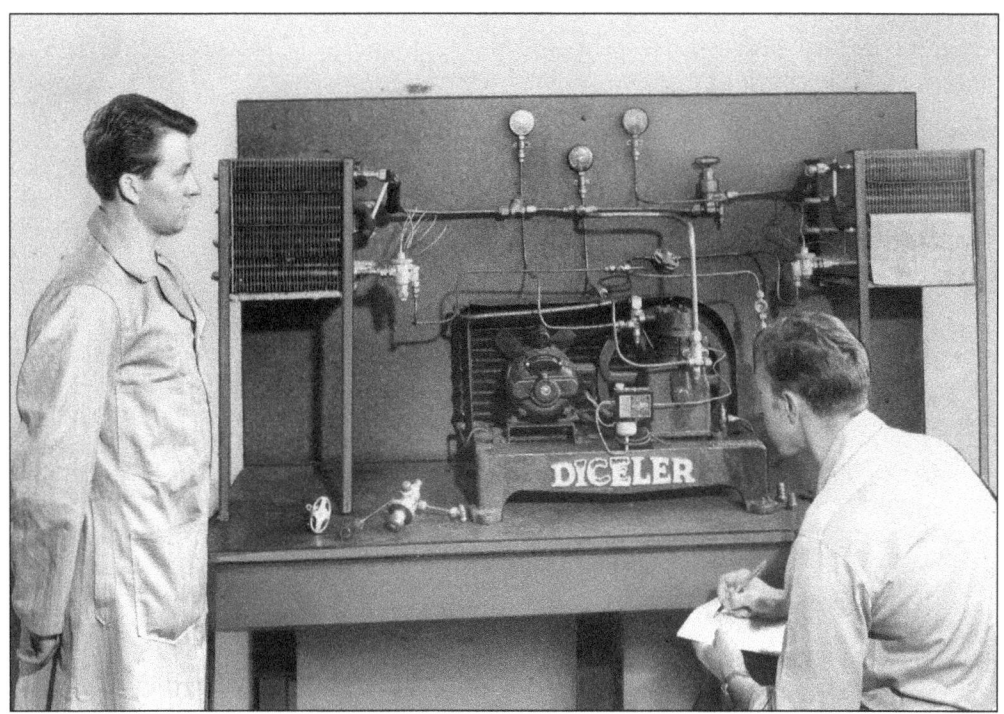

This photograph was also featured in the 1948–1949 *Photo Story*, and shows an instructor looking on as a student checks the performance of a two-temperature system employing a two-temperature constant pressure valve. Even as the school approached its 50th birthday, Oscar Werwath continued to place emphasis upon a student gaining practical familiarity with equipment and machinery that he may one day encounter in an employer's shop.

Despite all the attention paid to new technologies in the postwar years, the Milwaukee School of Engineering never neglected the foundation of electrical engineering from which it grew. Pictured here in 1946, three students are engaged in laboratory work with electric motors. The student on the right is checking the interior winding of a motor for short circuits. At center, a student is identifying the leads of the coil ends before connecting the armature to the commutator. This was done to prevent mistakes when he attached the delicate wires in place. The student on the left is hand winding a concentric coil. Techniques for producing small motors had improved during the previous 40 years, but Oscar Werwath steadfastly believed in the importance of "learning by doing" and required students to wind armatures by hand to familiarize themselves with the intricacies of the construction.

The war overshadowed much of the curriculum and industry partnerships made by the Milwaukee School of Engineering during this era, but that did not mean that student wellness was overlooked. Student participation in athletic activities and sports was high, and social gatherings with music, food, and drink were regular occurrences on the campus. In many of the promotional materials of the day, the school's Milwaukee location was celebrated as having close access to bathing beaches, skating lagoons, skiing facilities, and recreational parks, which all contributed to "healthful" living conditions for students. The photograph above features the school baseball team in the late 1930s, and a social gathering of students and instructors is shown below.

Oscar Werwath's unflagging dedication to his students is evidenced in this photograph, taken in 1946 during the reunion of early School of Engineering graduates from 1903 to 1911. These men would have eagerly attended the school to become electro-technicians and taken advantage of the budding field of electrical technology. Werwath can be seen seated in the center, holding a cane. Unfortunately, this would be the last time many of these men would encounter him alive. Oscar Werwath passed away on March 20, 1948. His death would shake the school to its core. For nearly 45 years, he worked tirelessly to make the Milwaukee School of Engineering into a unique institution through technical class offerings, partnerships with local industries, his enthusiasm for teaching, and his attention to the individual student. The loss was a powerful one, but many of his most valued interests and philosophies continue to resonate through the school today.

# *Four*

# MANMADE LIGHTNING
## EDUCATING THE PUBLIC

One of the traits that made Oscar Werwath so unique was his ability to teach anyone, regardless of skill level or aptitude. The Milwaukee School of Engineering was created because of the popularity of his technology lectures and his own love of learning. This trait shows through in his concept for the Concentric Curriculum, as well as his approach to hands-on learning. He was also attentive to leading scientific minds of the day and often brought them to the school for special lectures intended for students and later broadcast over the radio.

Students of the Milwaukee School of Engineering were not the only ones to benefit from Werwath's desire to teach. He also enjoyed teaching the public about his favorite subjects. Nowhere does this show through more clearly than in his creation of *Wonders of Electricity*—a traveling electrical show designed to both entertain and educate the public about scientific principles.

Even in the earliest years of his school, Werwath would hold public open houses where students displayed their work and demonstrated electrical experiments they had performed in class. As local industries began to take notice of his School of Engineering, Werwath rented exhibit space at trade shows to advertise the practical education offered at his school.

The first record of the electrical show appears in the 1917 school yearbook, where demonstrations with a Tesla coil and other electrical apparatus at the Milwaukee Motor Power Show are reviewed favorably. Photographs indicate that the marketing power of the event was not lost on Werwath; pennants and advertising for the School of Engineering festoon the stage.

*Wonders of Electricity* (also called *Wonders of Modern Electricity* and later the *Pageant of Electro-Magic*) was also performed at several notable events, including the 1934 Chicago World's Fair and the 1948 Centennial Wisconsin State Fair.

The show was an important part of marketing efforts well into the 1950s. Public education did not stop with the electrical show, however, as the school innovated new ways to blend teaching the public and promoting the values of the school. New technologies continue to support Oscar Werwath's dedication to educating the public even today.

In 1928, students of the School of Engineering attend a lecture entitled "Atoms as Wonder-Workers," given by Dr. Henry D. Hubbard, assistant director at the Bureau of Standards in Washington, DC, and creator of the Periodic Chart of the Atoms. Dr. Hubbard used models, apparatus, and lantern slides to ensure that his lecture was interesting and instructive.

He can be seen in the leftmost chair at right, facing the audience. School president Oscar Werwath is seated third from left in the seats facing the audience. Werwath invited many prominent figures in the scientific community to speak at his school for the benefit of the students.

Oscar Werwath encouraged his students to interact with the community to show off the cutting-edge technological work being done at his school. Given his desire to connect with local industry leaders as well as promote his school, it is no surprise that the School of Engineering had two exhibits at the Milwaukee Motor Power Show in March 1917. The first exhibit, pictured here, was a wireless station installed and supervised by the Wireless Club. This club had been organized by the School of Engineering on behalf of the local Boy Scouts. Student participants at the booth sent wireless messages for members of the public, demonstrated the equipment, and also chatted with any interested parties about the benefits of pursuing an education at the School of Engineering. As with many of Werwath's endeavors, this exhibition sought to educate and entertain as well as promote the school.

The second of the two exhibits set up by the School of Engineering at the 1917 Milwaukee Motor Power Show was the electrical exhibition. Students C.L. Chaffee, D.F. Trowbridge, and H.E. Clifton hosted an educational lecture and demonstration of electrical power to attendees. The lecture was given multiple times each day during the four-day exhibition. The hosts demonstrated manmade lightning using a Tesla coil, the extraordinary lifting power of an electromagnet, and the principles of an electric furnace. The 1917 school yearbook pronounced this demonstration "a brilliant success in every way." In later years, this show became a staple of the marketing efforts by the school, and the visually impressive lightning bolts from the Tesla coil appeared on countless promotional materials.

In this photograph from 1917, School of Engineering students C.L. Chaffee and D.F. Trowbridge can be seen on the stage of Milwaukee's Walker Hall with school president Oscar Werwath. A flier for this show called it "Elektrisches Wunderland" (Electric Wonderland), an acknowledgment of Milwaukee's German immigrant population and Werwath's own heritage.

This image from the 1930s shows one of the custom-made vans used to transport the show equipment throughout the central states. The exhibition schedule focused heavily on high schools, and a press packet suggested that in preparation for the show, schools should place posters in community shop windows, distribute handbills, make announcements to all classes, and arrange an automobile parade on the day of the presentation.

The experimental demonstration, referred to here as "Modern Electricity," was regularly performed on the School of Engineering of Milwaukee's campus on Thursday evenings during the 1930s by Oscar Werwath himself. It was performed in the Oneida Street (now Wells Street) building. The demonstration included such experiments as the decomposition of water, electro-heating with Nichrome wire, rapid freezing using dry ice, an electromagnet, electrical insulators, speech amplification, radio tubes, using a photoelectric cell to ring a bell, demonstrating how a television's scanning disc works, and of course, the Tesla coil. The photograph above shows the stage setup for the demonstration at the school, with Werwath at center. To the right is the cover of the informational booklet that was distributed to accompany the lecture.

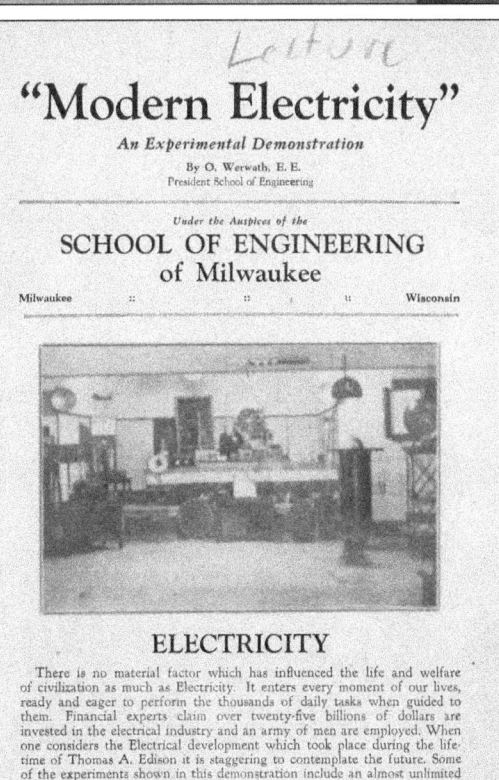

The traveling version of the *Wonders of Electricity* show was a brilliant success. Note that the School of Engineering of Milwaukee name is front and center, and a list of courses offered can be seen on the left side of the stage. Heralded as educational and informative, the show was also an invaluable way to promote the school.

Pictured here is a 1932 setup for *Wonders of Modern Electricity* at a Wisconsin high school. Although the display is pared down a bit from the stage shows for auditoriums, many of the same demonstrations were done, just on a smaller scale. High schools were especially important stops for the traveling show, as high school graduates were prime candidates for the Milwaukee School of Engineering.

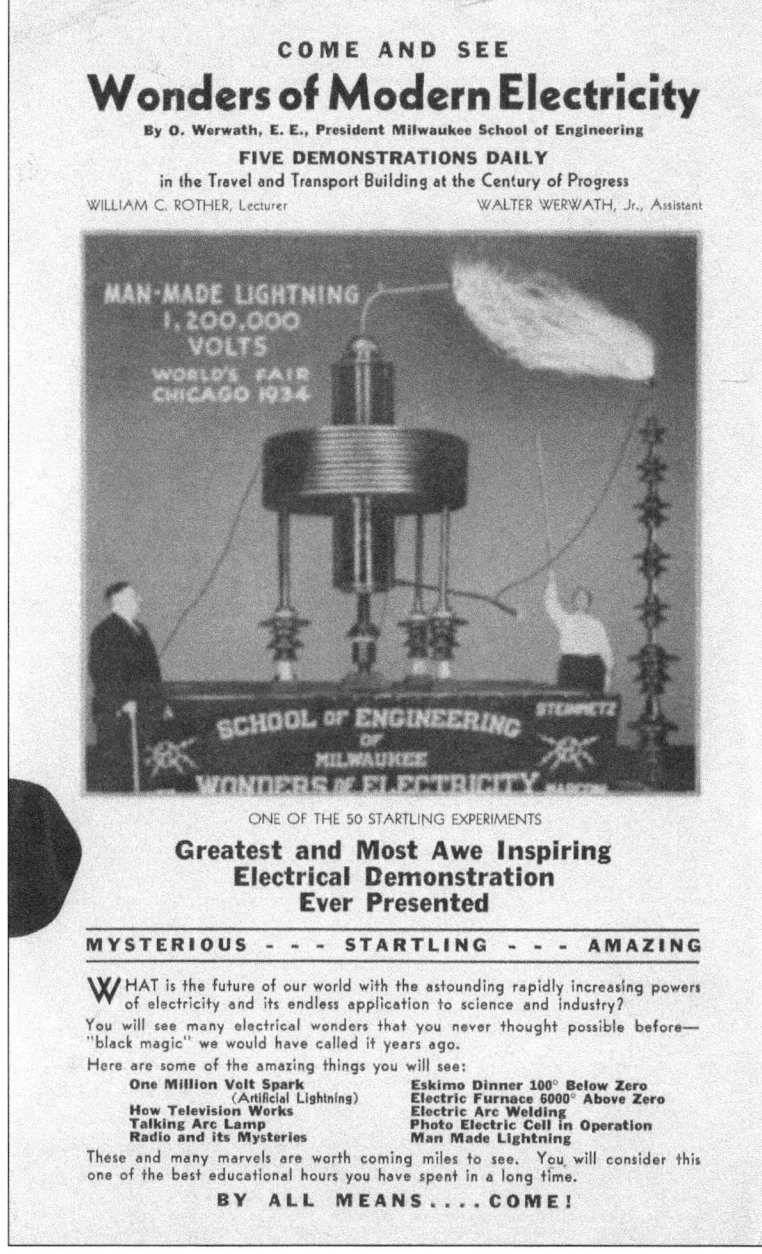

Oscar Werwath attended the Chicago World's Fair in 1934 and was so impressed by the displays of technology that he approached the fair's organizers about setting up the *Wonders of Modern Electricity* show at the fair. The organizers agreed, Werwath wrote the lecture script, and presenters William C. Rother and Walter Werwath Jr. headed down to Chicago to present the show five times daily to fair-goers. The popularity of this electrical show is a testament to Werwath's enthusiasm for technological marvels and his ability to demonstrate and explain complex scientific principles in a way that anyone could understand. His dedication to educating anyone who had an interest in science and technology made *Wonders of Modern Electricity* successful, both in terms of popularity as well as a marketing vehicle for his school. Seen here is one of the fliers given to attendees at the show.

This futuristic stage setting for *Wonders of Modern Electricity* was used briefly in 1941. The lecturer for this version of the show was Omar Larson. Many of the photographs taken during this particular lecture were used to advertise Milwaukee School of Engineering's new series of Industrial Defense Activities—day and night courses in welding, radio, and other skills that would prove useful during World War II.

The show was performed at notable southeastern Wisconsin events, including several consecutive years at the Milwaukee Midsummer Festival, 1946's Milwaukee Centurama, and pictured here, the 1948 Wisconsin Centennial State Fair. In this image, the presenter is demonstrating the strength of an electromagnet to a crowd of awed onlookers.

During the *Pageant of Electro-Magic* show from the 1950 Wisconsin State Fair, instructor Francis R. Drake and student Kenneth S. Jeszka demonstrate the explosive power of hydrogen after separating it out of water. Although the show had been evolving over the 30 years prior, many of the more startling demonstrations remained the same.

A student presenter of the *Pageant of Electro-Magic* demonstrates an experiment called "Uncontrolled Electrons," using the school's 500,000-volt Tesla coil. The presenter would insert the metal rod into the lightning discharge of the Tesla coil, and the fluorescent tube in his other hand would light without using wires for contact.

Never missing a chance to promote the Milwaukee School of Engineering to interested parties, the school also had an exhibit booth at the 1950 Wisconsin State Fair. The Concentric Curriculum and day courses can be seen prominently advertised on one side, and evening classes are displayed on the other. At far right is a sign promoting the *Pageant of Electro-Magic*.

A similar setup existed at the 1948 Wisconsin State Fair, with *Wonders of Modern Electricity* being performed on stage and a booth constructed nearby to promote the school. Here, Walter Werwath (one of the most prolific presenters of the electrical shows) and W.S. Harwood stand at the booth ready to answer questions from the public.

Pictured here is the most impressive Tesla coil owned by the Milwaukee School of Engineering, housed in an exhibition tent at the 1941 Milwaukee Midsummer Festival. This was the perfect venue for the electrical show, as much of the festival was dedicated to Milwaukee businesses showing off their products. Oscar Werwath was always looking for new ways to strengthen his school's ties to Milwaukee industries. Events like this exemplified his unique strategy for gaining notoriety for the school and provided varied avenues for marketing. The electrical show appealed to his abilities to simplify complex concepts and explain them to people of any skill level, and the association with local industries served to solidify the school's reputation as a great place to learn practical technical skills. Werwath's ability to appeal to laypeople and industry professionals alike was one of the character traits that made him so successful in his chosen field.

At certain events, the Milwaukee School of Engineering would perform the electrical demonstration and partner with an industry affiliate for booth space. Pictured here is the Milwaukee Electric Railway & Light Company's exhibit booth at the 1934 Chicago World's Fair. The company had been a supporter of the school for many years. There is a Milwaukee School of Engineering sign displayed, and school president Oscar Werwath poses at the front.

In this image from the mid-1930s, Oscar Werwath sits at an exhibit booth for Globe Battery. In the early days of the school, students assembled batteries in a workshop that would eventually be absorbed into the Globe Battery Company. Maintaining these long-term industry partnerships were and still are crucial to the school making sure that course offerings remain consistent with market needs.

As mass media on television and radio gained popularity, the second Milwaukee School of Engineering president, Karl O. Werwath (second from left), carried on with the new technology in the same spirit of his father's electricity demonstrations. Pictured here is a 1958 planning session for the weekly television special *Challenge*, on which Milwaukee School of Engineering faculty members interviewed leaders in industry about current issues in their field of expertise.

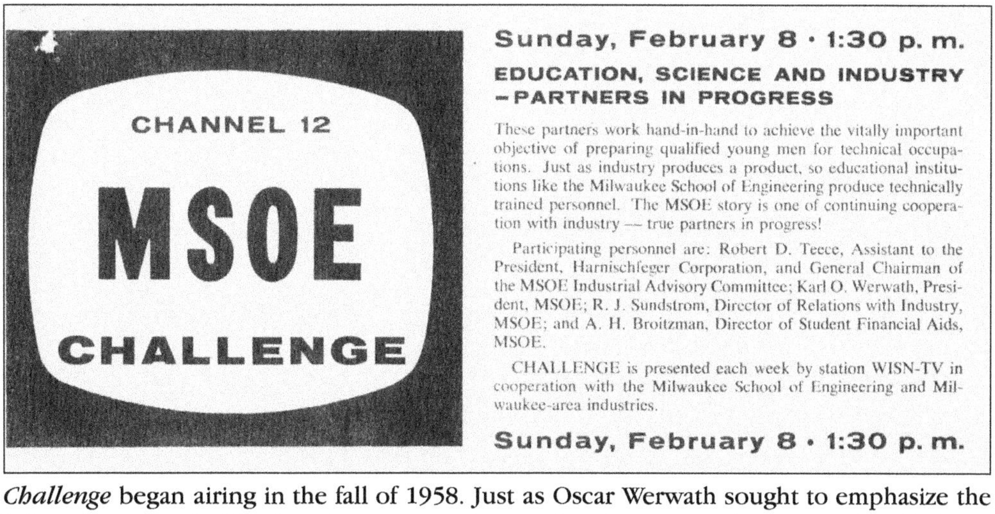

**Sunday, February 8 · 1:30 p. m.**
**EDUCATION, SCIENCE AND INDUSTRY**
**– PARTNERS IN PROGRESS**

These partners work hand-in-hand to achieve the vitally important objective of preparing qualified young men for technical occupations. Just as industry produces a product, so educational institutions like the Milwaukee School of Engineering produce technically trained personnel. The MSOE story is one of continuing cooperation with industry — true partners in progress!

Participating personnel are: Robert D. Teece, Assistant to the President, Harnischfeger Corporation, and General Chairman of the MSOE Industrial Advisory Committee; Karl O. Werwath, President, MSOE; R. J. Sundstrom, Director of Relations with Industry, MSOE; and A. H. Broitzman, Director of Student Financial Aids, MSOE.

CHALLENGE is presented each week by station WISN-TV in cooperation with the Milwaukee School of Engineering and Milwaukee-area industries.

**Sunday, February 8 · 1:30 p. m.**

*Challenge* began airing in the fall of 1958. Just as Oscar Werwath sought to emphasize the Milwaukee School of Engineering's connections with industry at tradeshows and exhibitions, Karl Werwath now achieved this via different mediums. Each *Challenge* episode was a half-hour in length. In 1963, the show was awarded a certificate of excellence by the Milwaukee Television and Radio Council. The show ran on Channel 12 until 1967.

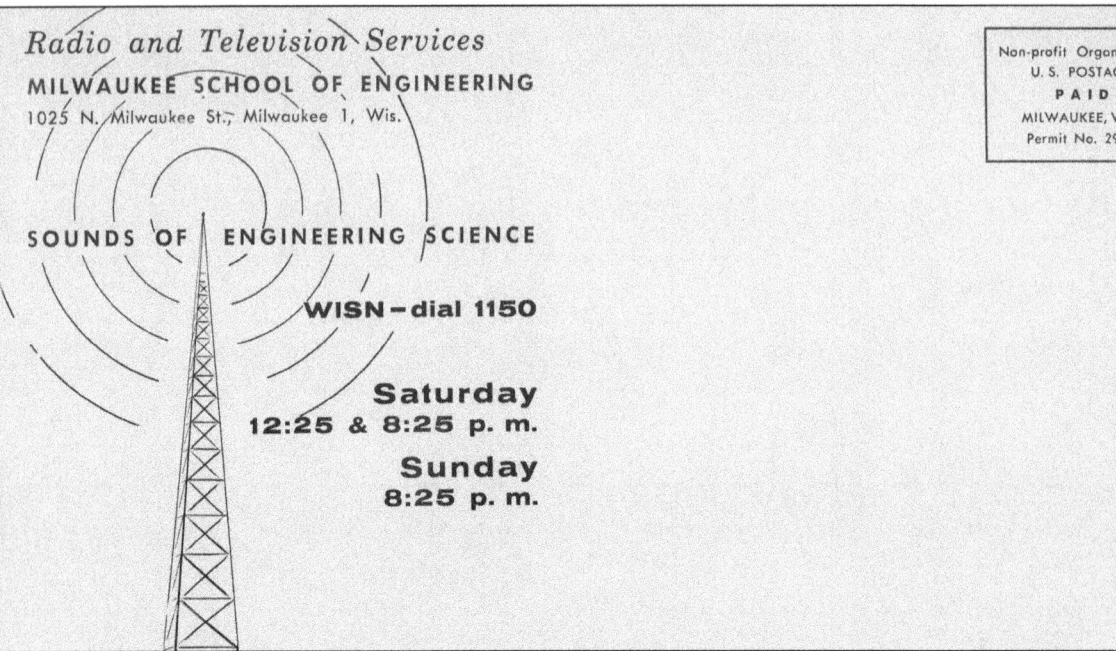

In addition to television, Karl Werwath also carried on his father's tradition of educating the public through a weekly radio program called *Sounds of Science*. The series ran from 1960 through 1977. Episodes were five minutes long and tackled technological topics in an uncomplicated manner. Episode titles included "Automatic Digital Computers," "Transistor Television is Here," "The Laser: A New Type of Light Source," and "Microwaves: The Future."

# Five

# RADIO WAVES
## RADIO AT THE SCHOOL

Radio and wireless technology was a passion of Oscar Werwath's from the start. Even before the School of Engineering existed, he spent his evenings demonstrating the inner workings of crystal radio sets to interested colleagues. Werwath was so intrigued by the progress made in radio technology during his lifetime that the second half of his "Forty Years of Achievement" address given to the Alumni Association in 1947 is a comprehensive look at radio developments at the school.

Werwath was especially proud of a wirelessly controlled zeppelin designed by School of Engineering students that amazed crowds at the 1911 Merchant and Manufacturers Association exhibition. Students at the school were encouraged to experiment with radio equipment, and soon after the zeppelin, a wirelessly controlled monoplane was designed by another student.

The School of Engineering of Milwaukee operated one of the first three radio broadcasting stations in the state of Wisconsin in the early 1920s. After an upgrade in equipment, the station's call numbers changed from WIAO to WSOE. This station gained significant popularity in Milwaukee, as it hosted a variety of live music performances, educational lectures, and scheduled studio musicians. The station was open to the public, and tours of its equipment and studios were given by student volunteers. WSOE was sold in 1926, and its equipment was put to use for station WISN.

Several groups of radio enthusiasts formed clubs at the Milwaukee School of Engineering through the years. The Amateur Radio Club was born in 1926 and operated under the call numbers W9SO. This club achieved the first radio contact with hurricane-devastated Cuba in 1926, and communicated with a South Pole expedition. Radio classes were formally added to the school's curriculum in 1933, and the club dissolved. In 1949, the Amateur Radio Club was reborn under the call numbers W9HHX.

A donation from alumnus Everett Cobb allowed station WMSE to begin broadcasting on March 17, 1981. A local favorite, WMSE prides itself on eschewing traditional radio and providing Milwaukee listeners with a wide variety of music and programming.

There is no doubt that Oscar Werwath would be pleased by the tradition of radio continuing at his school.

Before the Milwaukee School of Engineering was wowing audiences with its Tesla coil, the school astonished the 1911 Merchant and Manufacturers Association with a wirelessly controlled zeppelin built by students. Each wireless flight demonstration would conclude with the zeppelin unfurling a flag and a phonograph on board striking up a patriotic air. Several reviews of the event called this one of the most outstanding attractions at the industrial exhibition.

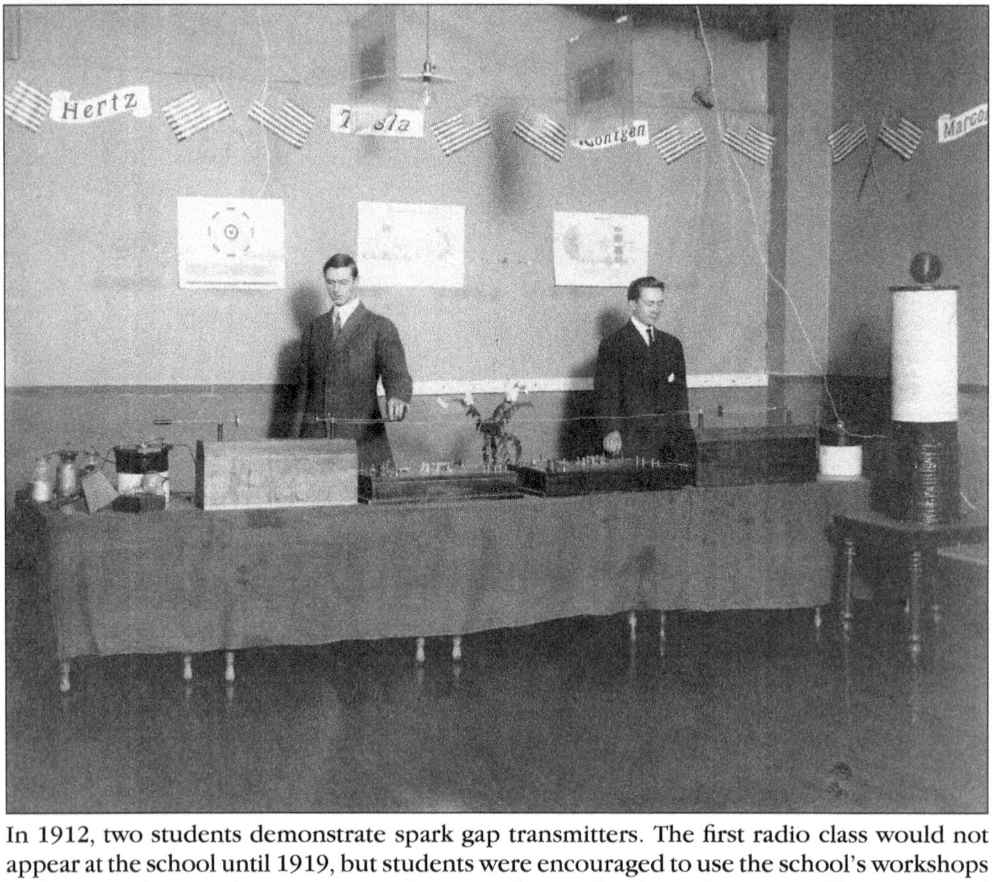

In 1912, two students demonstrate spark gap transmitters. The first radio class would not appear at the school until 1919, but students were encouraged to use the school's workshops to design and assemble equipment relevant to their technological interests. There is a good chance that these transmitters were student-built.

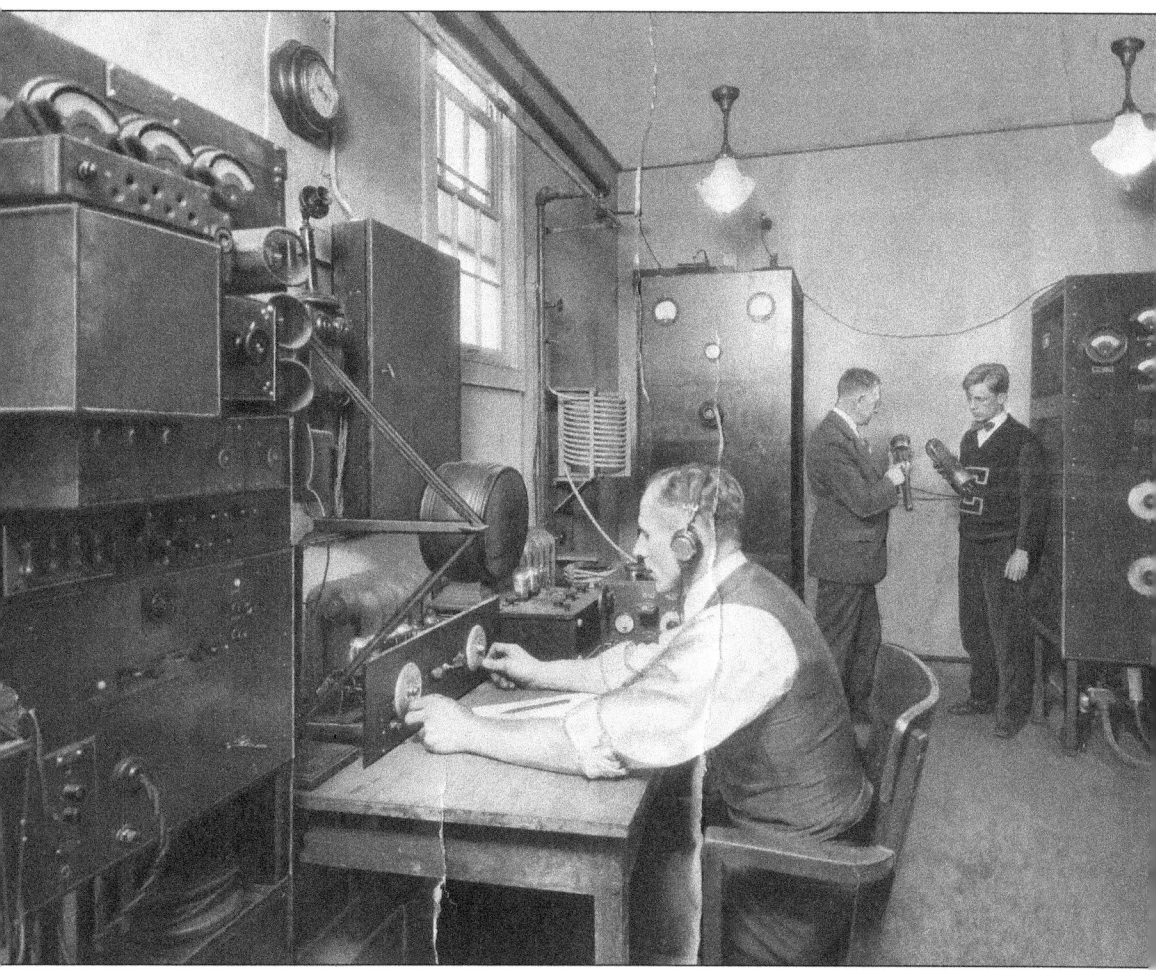

Seen here is the operating room in the Marshall Street building for School of Engineering of Milwaukee radio station WIAO in 1922. WIAO is said to be one of the first three radio stations in the state of Wisconsin. The power for the station was derived from two enormous sets of storage batteries, each having 800 cells in series and capable of furnishing a plate potential of 1,600 volts. One battery was used at a time, and the other would remain charged in case of an emergency. A three-stage impedance coupled circuit served as the amplifying equipment, and the station itself was rated at 250 watts. Two tubes were used as oscillators while two more were used as modulators, and the station used a piezo crystal oscillator to stay on its assigned frequency. WIAO was fairly short-lived, and the name was changed to WSOE in 1925 when more powerful equipment was installed.

WSOE was the first radio endeavor by the School of Engineering of Milwaukee to receive significant public recognition. It was known in the Milwaukee area for inviting music groups to the Ensemble Studio at the station to perform for the radio audience. Pictured here in 1926, an announcer for WSOE introduces the Manhattan Orioles, who performed live on-air.

Pictured here is the Solo Studio at the WSOE headquarters in the Marshall Street building in the 1920s. This room is where staff soloists would perform single numbers, duets, and small group selections. The station also had an Ensemble Studio, which contained two pianos, lots of seating, and a chorus stage. Both studios had windows looking into the reception room, where listeners could request their favorite numbers.

In 1926, the most visible pieces of WSOE equipment were the two 150-foot towers on either side of the Marshall Street building, which supported a four-wire T-antenna and a six-wire T-counterpoise, used with a newly designed tank circuit. The studio, transmission, and battery rooms were located on the main floor of the building and were open to the public.

Pictured here is the WSOE transmitter room in the Marshall Street building in 1926. Only a few years earlier, the School of Engineering of Milwaukee inaugurated its first series of radio engineering classes. Due in part to the popularity of stations like WSOE, these courses taught students to design, install, construct, inspect, and test radio equipment.

The school's Amateur Radio Club formed in response to federal laws prohibiting schools from commercializing over the radio and followed the sale of WSOE in 1926. Students interested in radio broadcasting began amateur station W9SO, which was operated by club members. Pictured here is the club in 1927; from left to right are Rueben Tayek, Norb Richard, William P. Gainer, and Harold Homberg.

Amateur radio station W9SO was chosen to be the College Radio Union's key radio station for the western section. This meant that W9SO had sufficient power and schedule time to record the reports of smaller stations and reported up to the College Radio Union in Washington. In 1927, Norb Richard (left) and William P. Gainer are pictured monitoring the airwaves from their club room on the roof of the school building.

By 1928, the Amateur Radio Club was growing and gaining popularity among students. Within its first seven years, the club had gained notoriety by being the first station to report information on a devastating hurricane in Cuba in 1926 and communicating with Admiral Byrd's ship on its exploration of the South Pole. Note the QSL cards hung on the walls.

QSL cards were (and to some extent still are) a way for an amateur radio station to confirm reception of another station's signal. Sent through the mail like a postcard, generally these cards include information about time received, location, and frequency. This is a 1928 QSL card sent by William Gainer to previous Amateur Radio Club president Ben J. Chromy after he moved his ham station to Washington, DC.

Pictured here is the Amateur Radio Club's trip by boat to inspect the equipment at a Chicago radio station and attend the 1930 Chicago Radio Convention. Inspection trips were an important part of education at the Milwaukee School of Engineering and were certainly not limited to future radio technicians. During each term, one entire week was dedicated to inspection trips to local industries—power plants, manufacturing companies, and the like. Efforts such as this served to get students out into the larger community and encouraged them to interact with professionals in their chosen field of study. Just as Oscar Werwath and his students educated the general public through radio lectures and the weekly *Sounds of Science* program, industry leaders were encouraged to educate students about current issues in the field. Only a few years later, for the 1932–1933 school year, courses in radio instruction would be changed from a handful of evening classes to a dedicated two-year program.

The new radio classes diminished the need for a formal radio club, and in the years between 1933 and 1949, the club dissolved. It was not until 1949 that the Milwaukee School of Engineering Amateur Radio Club formed. Pictured here, a radio operator for amateur station W9HHX demonstrates the use of radio equipment to interested onlookers.

The Amateur Radio Club received its call numbers from the Federal Communications Commission on May 19, 1949, and in July of the same year, it was granted a charter by the American Radio Relay League. The goals of the club were simple: to foster goodwill among everyone interested in amateur radio and provide a facility at the school for interested parties to obtain licenses.

Broadcasting and receiving wireless transmissions had been an interest of students at the Milwaukee School of Engineering since the very beginning. The same can be said for the building and servicing of broadcasting equipment. Reminiscent of the time when Oscar Werwath lectured to his colleagues and demonstrated the workings of crystal radios on his kitchen table, building and servicing radio equipment was also an important subject at the school. A formal radio broadcasting curriculum was not implemented until 1932, but building radio equipment was a crucial part of the practical electrical education provided. Here, students in 1926 are testing and adjusting radio receiving sets in the school's only Radio Laboratory. They were likely part of a special three-month course in radio, which focused on designing, merchandising, installing, and servicing radio equipment.

In 1935, a one-year course in radio and television engineering was introduced at the school. Following Oscar Werwath's Concentric Curriculum, this new program gave students interested in the subject of radio two course paths to choose from—equipment servicing or broadcasting. These students are being drilled on Morse code to ensure that they can meet the required speeds.

Because of an influx of self-styled radio technicians overstating their skills to employers, a system of examination was implemented by the Wisconsin Radio Trade Association. Class C had little practical experience and was cleared only to do simple installations, Class B could do general service work, and Class A could manage a radio department and construct equipment. At the Milwaukee School of Engineering, Class A graduates were the goal.

Television broadcasting came to Milwaukee in 1947. In true Milwaukee School of Engineering fashion, television servicing and repair was added to the already-established radio program in 1948. The program was expanded from a single year to 18 months to accommodate the new material. Television classes focused on the servicing of receivers, the operating of transmitters, control room and studio procedures, and ultra-high frequency techniques. Students who completed this 18-month technician course could also continue for an additional two years and receive a bachelor of science degree in electrical engineering with a major in electronics. Pictured here are two views of the Television Laboratory in 1950. Students are performing laboratory exercises localizing faults and determining defective components in receivers.

The Evening Division of the Milwaukee School of Engineering was re-opened in September 1949. Classes in technical specialties were available through the Institute of Electrotechnics, the Refrigeration, Heating and Air Conditioning Institute, the Welding Institute, the Department of Mechanics, and the Department of Drafting. The more intense electrical subjects the school was known for, such as television, consisted of 10 months of evening classes. The image above shows students from 1950 following laboratory instructions as they test the inner workings of a television set. At right, also in 1950, students practice servicing a television set.

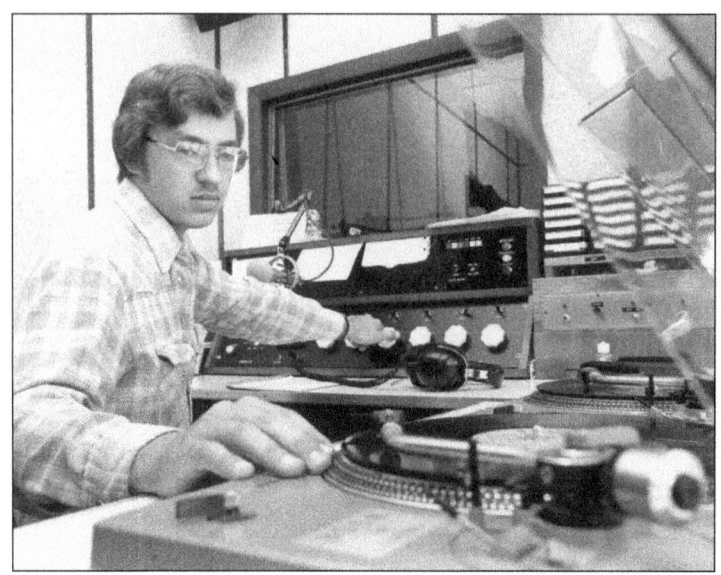

A far cry from the wireless spark signals sent by students at the start of the 20th century, Milwaukee School of Engineering radio station WMSE has been broadcasting alternative and freeform music over the airwaves since 1981. In this photograph, WMSE disc jockey Paul Pergandi plays a record from the broadcasting booth in Margaret Loock Hall.

Radio signals have been a part of the Milwaukee School of Engineering since the days when Oscar Werwath gave lectures about crystal radios around his kitchen table. The technology used to broadcast content over the airwaves may have changed substantially, but that same passion for using new technologies is alive and well at the school. Pictured here, two WMSE disc jockeys pose outside the studio in the early 1980s.

# *Six*

# SUSTAINABLE POWER
## TECHNICAL EDUCATION CONTINUES

When Milwaukee School of Engineering founder and president Oscar Werwath passed away in 1948, his eldest son, Karl, was elected by the Board of Regents to take the reins. As a young man, Karl had always been a part of the campus, whether he was helping to maintain the equipment rooms, teaching classes in economics, or performing the duties of vice president. Well-educated in electrical engineering, teaching, economics, and business, he was the perfect candidate to take over the job.

The first few years of Karl Werwath's presidency took a decidedly introspective tone. Many of his contributions to the school during this time were made behind the scenes, at the administrative and advisory level. He formed a number of committees to conduct long-term studies on nearly every aspect of the school. These studies would help him to make informed decisions in the years to come.

Regarding decisions about matters of the school curriculum, Karl Werwath continued to rely on the expertise of his father's industrial advisors to ensure that class teachings remained aligned with the needs of industry. Class offerings remained largely unchanged during this period, with the newest courses being in refrigeration, heating, ventilating, and air conditioning. These courses were offered in increments depending on a student's desired position, similar to Oscar Werwath's Concentric Curriculum.

As he searched for his own presidential goals for the school, Karl Werwath continued to support the traditions and ideologies that his father had kindled to make the school occupy such a unique space in the educational climate of Milwaukee. He made sure that course instruction included elements of practical education, especially trying to simulate those conditions that a student might encounter in the field. He also kept the less scholarly, but more fun tradition of the campus Saint Patrick's Day celebration alive.

Karl Werwath's first few years as president of the Milwaukee School of Engineering were calm, and enrollments were stagnant as a result of the Korean War. Although most of his work during this time was through research and self-study, he was just preparing to begin building his legacy in the years to follow.

Karl O. Werwath was born on March 1, 1909, to Oscar and Johanna Werwath just six years after the School of Engineering opened. He had three younger siblings, Greta, Hannah, and Heinz. This photograph was taken on the day of his graduation from the Milwaukee University School in 1927.

Karl Werwath shared his father's passion for all things electrical, and in 1936, he graduated from the Milwaukee School of Engineering with a bachelor of science degree in electrical engineering. He continued his education at the Northwestern School of Education Extension in educational administration and the University of Wisconsin for economics and business administration.

Karl Werwath was always involved in the Milwaukee School of Engineering, working in the stockroom as a young man, the admissions department during his time as a student, an instructor in economics, and eventually as vice president. When his father passed away in 1948, Karl's education and experience at the school made him the best candidate to take Oscar's place as president. The Board of Regents unanimously voted him into the position in April 1948. His brother, Heinz Werwath, was treasurer of the school and had spent time working in the admissions department, as assistant registrar, and as director of admissions. Here, the two brothers (Karl at left) sit at a table beneath a painting of their late father donated by the Associated Alumni Endowment Foundation. As the school still reeled from his father's passing, Karl would struggle to fill his shoes.

For the most part, the coursework at the Milwaukee School of Engineering remained the same after Oscar Werwath's passing. The emphasis on electrical engineering and the importance of a practical education were still prominent, as evidenced by these young men participating in the Practical Electricity Laboratory in 1950.

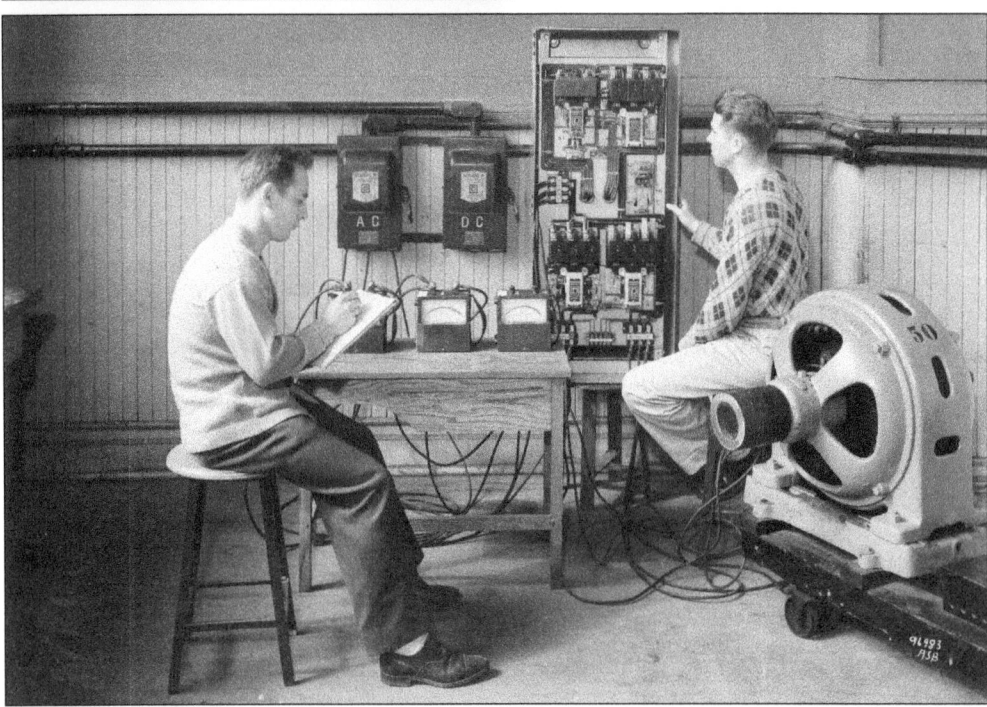

In 1950, two students are recording the results of a test using a dynamo and a series of meters. Equipment stations were arranged for convenience and also to simulate the conditions students would face when they entered the workforce. Hailing back to Oscar Werwath's laboratory arrangements, large equipment like motors and generators were attached to boards with wheels so they could easily be moved around the laboratory.

In the early 1950s, the Institute of Electrotechnics focused on electricity, power, radio, and television. Here, students from this branch of the school are shown installing electrical outlets, switches, and control units in conditions that simulate those found in a newly constructed building. The skills learned during the first term of the program were designed to allow students to move seamlessly into more advanced electrical topics.

Pictured here in 1960 is a scene in the Milwaukee School of Engineering Electrical Technology and Engineering Laboratories. Two students are working to take down the results of an experiment using a cathode ray oscilloscope. Students were encouraged to work in pairs or small groups when completing their laboratory experiments.

Just before his death in 1948, Oscar Werwath made some significant changes to the Milwaukee School of Engineering campus. He had added a new building to house the refrigeration laboratories and built new physics and chemistry laboratories and classroom space. He also added many new subjects to the Evening Division, bringing the numbers up to 12 subjects offered in 15 departments. This was the first time the Evening Division had offered courses leading to a technician's certificate in electronics. Seen here are two images of classrooms from this time—above from a math lecture room and the below from a drafting room. Although the school was shaken to its core by the passing of Oscar, the foundation of education remained the same. It would take Karl Werwath a few years before he began to make his mark on the campus.

Partnerships with local industries remained important to the Milwaukee School of Engineering, and Karl Werwath knew the value of maintaining these mutually beneficial relationships. The Materials Testing Laboratory was a place where students and industry leaders would work together to find solutions to real-world challenges. Such an arrangement was a true innovation, with its roots in Oscar Werwath's earn-while-you-learn program, student inspection visits to local industries, and lecture sessions with visiting business leaders. Pictured here are students at work in the Materials Testing Laboratory in the 1950s. In the coming years, this laboratory would become a crucial part of Karl's legacy at the school.

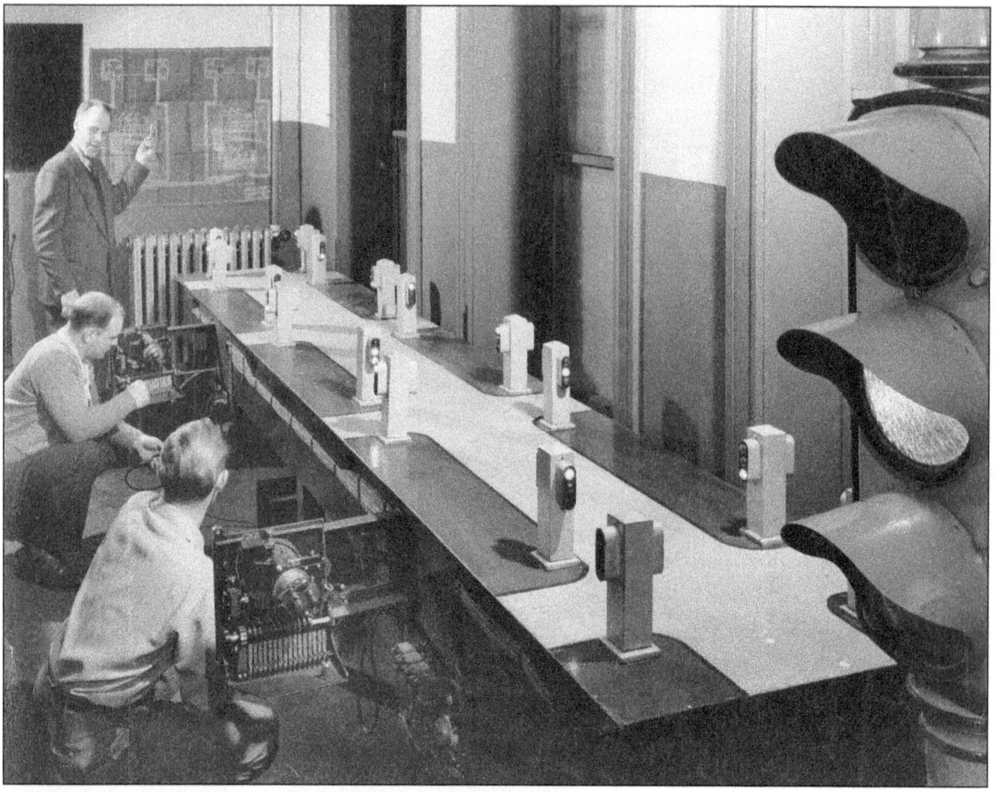

In 1950, students in Rueben Sheflog's practical electricity class use a model traffic signal system to conduct experiments. They are checking for continuity in the circuits and learning how the various mechanisms of the traffic signals operate. Sheflog stands at the back with a blueprint of the setup.

Practical electricity education took on many forms at the Milwaukee School of Engineering, from building storage batteries, to house wiring, to radios and television. Here, two students of practical electricity compare the parts of a traffic signal with blueprints in the interest of acquainting themselves with all parts of the device.

In September 1950, the Milwaukee School of Engineering offered a special weeklong welding course for women during an academic break. Nine women from the Heil Company participated in the class, which was held in the Welding Shop. In October, the school bulletin noted that the event "practically makes MSOE co-educational."

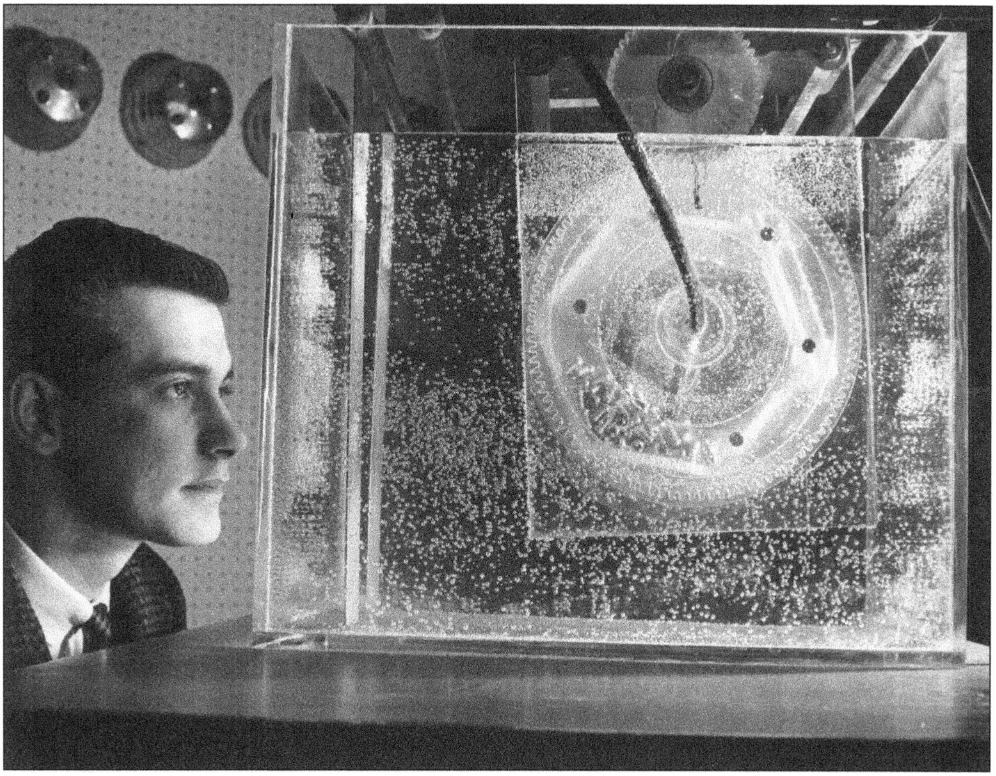

Open houses for the public and prospective students continued during Karl Werwath's presidency. These events allowed industry partners and other members of the community to learn about the facilities and the institution's academic excellence. In this image from 1961, an attendee at one of these open house events views an electroplating exhibit in the Metallurgical Laboratory.

Along with his father's focus on electrical education, Karl Werwath also maintained the more lighthearted traditions at the Milwaukee School of Engineering. Celebrating Saint Patrick on campus dates to 1934, when the student branch of the American Institute of Electrical Engineers dedicated itself to the proclamation, "St. Patrick was a Hell of a Good Engineer." Pictured here, Werwath symbolically relinquishes his presidential power for a single day to a costumed Saint Patrick.

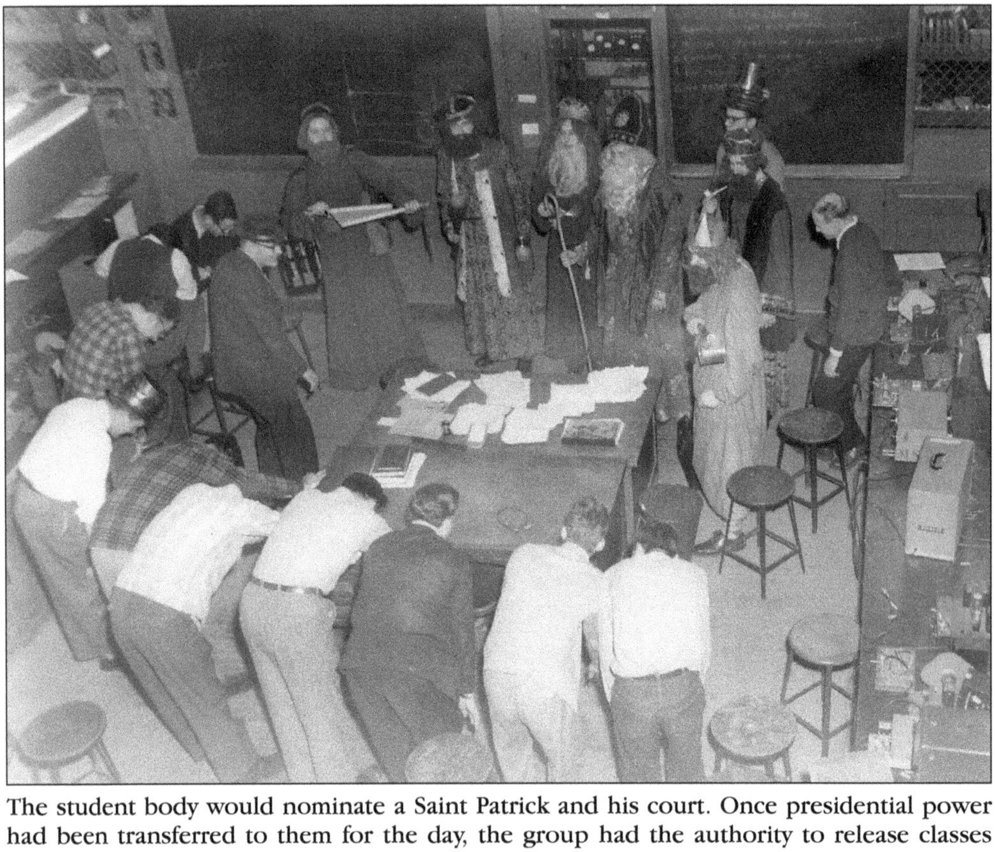

The student body would nominate a Saint Patrick and his court. Once presidential power had been transferred to them for the day, the group had the authority to release classes at will and generally wreak havoc on campus. Often, they would dress in false beards and outlandish disguises during the celebration. Featured here is the Saint Patrick's Day court from 1952.

Pulling pranks such as cutting off the ends of instructors' ties and disrupting classes abounded during the Saint Patrick's Day celebrations. The honorary Saint Patrick could also make decrees that had to be followed, such as wearing one's jacket backward for the remainder of the day. In this photograph of the closing dinner of the 1949 festivities, several students can be seen adhering to Saint Patrick's wishes.

The Saint Patrick's Day tradition continued through Karl Werwath's presidency and is still celebrated on the Milwaukee School of Engineering campus today. In this image from the late 1970s, then-president Robert Spitzer prepares to relinquish his presidential power to Saint Patrick and his court. Today, the tradition has evolved with the school, but the spirit of Saint Patrick's Day remains the same.

The acquisition of the Milwaukee Street Building in 1948 marked the change in presidencies as Karl Werwath took the reins from his father. Enrollment flagged a bit during this time, as the United States became involved in the Korean War. These early years of the second Werwath presidency are marked mostly by the implementation of long-term studies, the effects of which would not be seen until much later. Karl Werwath worked hard to gain and maintain relationships with donors and industry leaders, and formed industrial advisory committees for each of the school's major course tracks. With the help of the industrial advisory committee and the Board of Regents, he was able to offer mechanical engineering classes beginning in the fall of 1952. As the son of the school's founder, Karl had the legacy of an innovator to live up to. After a few years conducting careful research and conferring with his advisors, he was ready to begin constructing his legacy.

# *Seven*

# BUILDING A CHARGE
## GROWTH AT THE SCHOOL

The latter half of the 1950s mark the beginning of Karl Werwath's efforts to see his grand plan for the campus come to fruition. His Milwaukee School of Engineering Corporation Self-Study Committee recommended that a new development program should be set up for the school. The goal of this plan was to build the "MSOE of Tomorrow." Content with the blessing of his advisors, Karl Werwath set about making his grand vision, entitled Technology Park, a reality.

Two new buildings were needed to meet the growing enrollment numbers. First came the Allen-Bradley Hall of Science in 1960. It was the first building acquired by the school with the specific intention of completely redesigning it into technological education facilities. The second building was opened in 1967. The Fred Loock Engineering Center boasted increased floor space for laboratories and shops, along with ultra-modern equipment.

Space for educational endeavors had been increased, but students did not have large dormitories or housing options available on campus. They did not have to wait long. The Roy W. Johnson Residence Hall opened in 1965, and the Margaret Loock Residence Hall opened in 1967.

The importance Oscar Werwath placed on maintaining and nurturing relationships with industrial partners was certainly not lost on Karl. The school acquired the Leather Institute in 1960, which conducted research on leather tanning techniques. Next came the Fluid Power Laboratory, occupying 12,000 square feet in the Fred Loock Engineering Center in 1962, and intended to provide the opportunity for both academic and company-sponsored research in fluid power. In 1967, the campus also received the enormous 25 million electron volt Betatron, donated by the Allis-Chalmers Company, with accompanying traveling cranes provided by the Harnischfeger Corporation.

Karl Werwath made it a goal to provide courses using the newest technologies for his students' education, just like his father. In 1958, the first computer technology curriculum was added to the Milwaukee School of Engineering.

The legacy Karl Werwath left behind on campus was one of improved physical spaces, which proved invaluable as the school continued its growth. Karl Werwath served as president until his retirement in 1977.

The start of Karl Werwath's true legacy at the Milwaukee School of Engineering began with the opening of the Allen-Bradley Hall of Science in May 1960. After being acquired by the school, it had undergone a complete redesign and was updated to house educational facilities. Touted as "ultra-modern," the building served as the cornerstone for Werwath's grand Technology Park objective.

In 1965, ground was broken on the next addition to the school, the Fred Loock Engineering Center. The building was opened in 1967, and offered modern classrooms and laboratories specially designed for the school's needs. It housed the specialized equipment necessary for the study of air conditioning, fluid power, industrial engineering, internal combustion, metallurgy, and welding.

Two new student dormitories were added to the Milwaukee School of Engineering at about the same time. The Roy W. Johnson Residence Hall, which opened in 1965, and the Margaret Loock Residence Hall, which opened in 1967. The Roy W. Johnson Residence Hall was created with the idea that a student must have a place designed for both studying and exchanging ideas with his peers. Margaret Loock Hall was built to house 429 students and had laundry facilities, meeting rooms, and a visitor's lounge. Above, Karl Werwath (third from left) and Fred Loock (center) admire plans for the new Margaret Loock Residence Hall in 1964, along with Milwaukee mayor Henry Maier. The photograph to the right shows the two completed residence halls: Roy W. Johnson Residence Hall at left and Margaret Look Residence Hall at right.

During this period of the Milwaukee School of Engineering's history, many plans to expand the campus space were made, but not all came to fruition. Pictured here in 1972, regent Erhardt C. Koerper (left) and Heinz Werwath pose in front of a campus design that would not get past the planning stages.

By the 1960s, the Milwaukee School of Engineering library had evolved from a couple of shelves of donated books in the English department to a space in the Milwaukee Street Building. It had two professional librarians on staff and boasted over 30,000 volumes. The collections would continue to expand in the coming years, and in 1980, the Walter Schroeder Library opened with a dedication featuring former US president Gerald Ford.

One of Karl Werwath's magnificent visions for the Milwaukee School of Engineering campus was Technology Park. The idea so impressed the City of Milwaukee, and it saw what benefits could be brought to the city by a large, technologically focused university, that it set aside additional space near the school for future development in the late 1950s. Further planning for the expansion continued, and in 1957, the Milwaukee School of Engineering Corporation Self-Study Committee recommended preparing a development program for the school. The end goal of this endeavor was called the "MSOE of Tomorrow." The addition of the Allen-Bradley Hall of Science and the Fred Loock Engineering Center to the campus helped Werwath add weight to his vision and persuade other potential donors to get excited about the futuristic new direction of the school. Here, a model of his Technology Park is seen behind him. Its footprint was very similar to that occupied by the school today.

Pictured here is a student performing an experiment using a carbon-sulfur determinator in the late 1950s. This piece of equipment was located in the Metals Laboratory ("M") Building on the Milwaukee School of Engineering campus. When it was added to the campus in 1957, the building added an additional 25,000 square feet of laboratory space to the school.

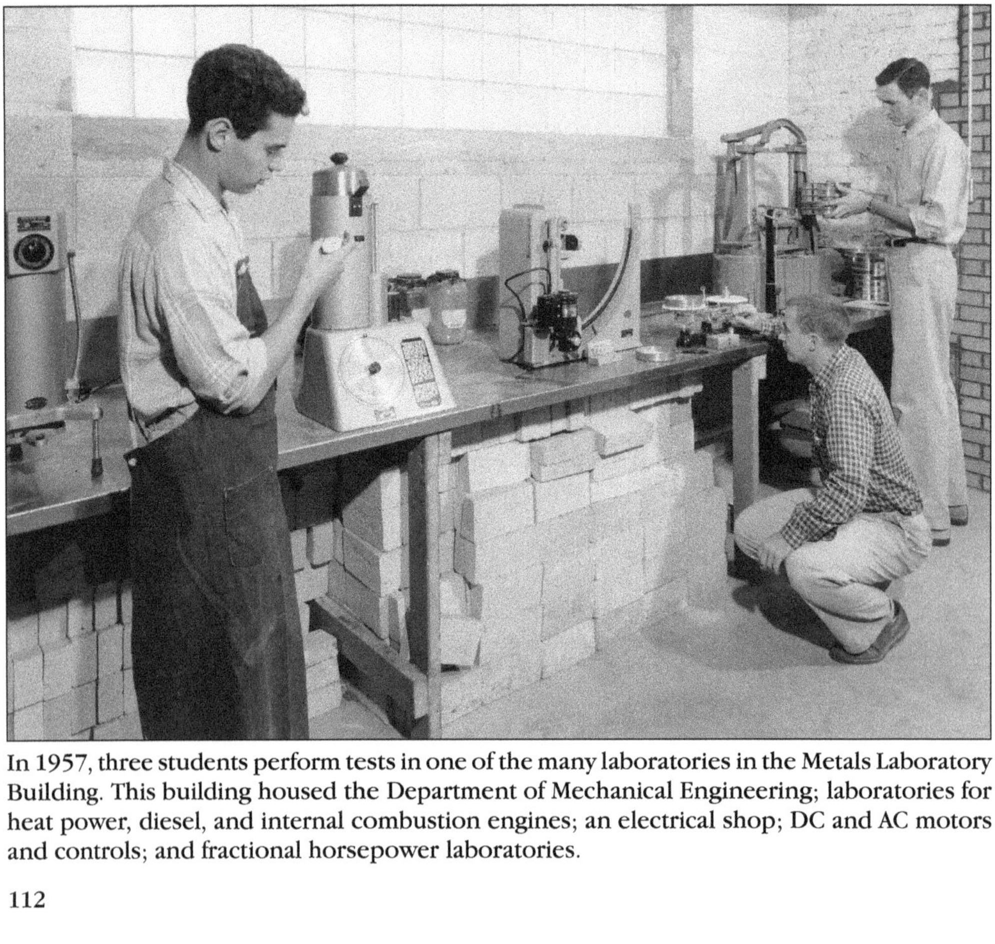

In 1957, three students perform tests in one of the many laboratories in the Metals Laboratory Building. This building housed the Department of Mechanical Engineering; laboratories for heat power, diesel, and internal combustion engines; an electrical shop; DC and AC motors and controls; and fractional horsepower laboratories.

The Fluid Power Laboratory was established on the Milwaukee School of Engineering campus in 1962 with the goal of providing education, applied research, and technical publications in fluid power technology. The institute was overseen by a governing board staffed with members of industry partners with an interest in fluid power research. By 1967, the institute was recognized by the National Fluid Power Association, which established the Fluid Power Hall of Fame at the school. Both images from the 1960s show students interacting with the specialized equipment necessary for the study of fluid power dynamics. Today, the Fluid Power Laboratory has evolved into the Fluid Power Institute, a national leader in fluid power research, with a focus on hydraulic fluids.

The Leather Institute moved to the Milwaukee School of Engineering in 1960, after spending 25 years at Lehigh University. It was housed in the Allen-Bradley Hall of Science; like the Fluid Power Institute, it was overseen by a managing body of industry leaders interested in researching leather processes. Its primary purpose was to study the fundamental leather processes for the six leading leather-producing companies and to find solutions for pollution control of industrial wastes. The mission of the institute was to publish research papers, disseminate applied technical information to related industries, establish a noncredit course in basic leather technology, and actively participate in national societies of the leather industry. In 1966, the institute was celebrated for developing a new method of manufacturing high-quality shoes from leather. In both of these images from 1961, leather technicians at the institute work in the laboratories.

In 1971, research activities at the Leather Institute were phased out by sponsors of the Midwest tanneries and chemical companies focused on the production of chromium. The institute was completely deactivated, and was replaced with the Environmental and Water Resources Laboratory. In part, the Leather Institute had focused on water resources and potential ways to prevent industrial waste from reaching the general water supply, so much of the equipment was the same. In October 1971, the Milwaukee School of Engineering Alumni Association sponsored a symposium on "The Engineering Challenge in Environmental Control" to celebrate the evolution of the lab and meet with industry leaders with experience in the field. It was a main feature of the 1971 homecoming activities and was an excellent way to make partners in industry aware of the new direction for Milwaukee School of Engineering research. Pictured here, laboratory supervisor Dr. Eugene Magnuson examines some laboratory results.

Of all the enhancements to the Milwaukee School of Engineering campus during Karl Werwath's presidency, one of his proudest was the addition of a 25 million electron volt Betatron. The specialized piece of equipment was donated to the Milwaukee School of Engineering by the Allis-Chalmers Company. Two large traveling cranes for the setup were also donated by the Harnischfeger Corporation. This piece of equipment was so large that requirements for its housing were specially considered during the building stages of the Fred Loock Engineering Center. It was given a dedicated space three stories high, with thick concrete walls to contain any radiation. The Betatron was housed in the Institute for Nondestructive Testing, which was soon renamed the Radiography Laboratory. The Betatron was intended to help members of the institute solve unique nondestructive testing problems for industry. From left to right, president Karl Werwath, Willis G. Scholl (Allis-Chalmers), Walter Harnischfeger (Harnischfeger Corporation), Fred Loock (regent chairman), and Heinz Werwath (vice president) participate in the 1967 unveiling and ribbon cutting ceremony.

These two images show the scale of the Betatron equipment in its housing in the Fred Loock Engineering center. The photograph at right was taken after its unveiling in 1967, and the image below shows a student using the equipment in 1970. The Betatron was an electron accelerator type of x-ray unit that was used for industrial radiography. It was intended to inspect heavy weldments and castings, as its high-energy x-ray beam could penetrate massive steel objects up to 20 inches thick. Because of the relative rarity of this type of equipment, the Institute for Nondestructive Testing was made available to industrial firms for research and radiography.

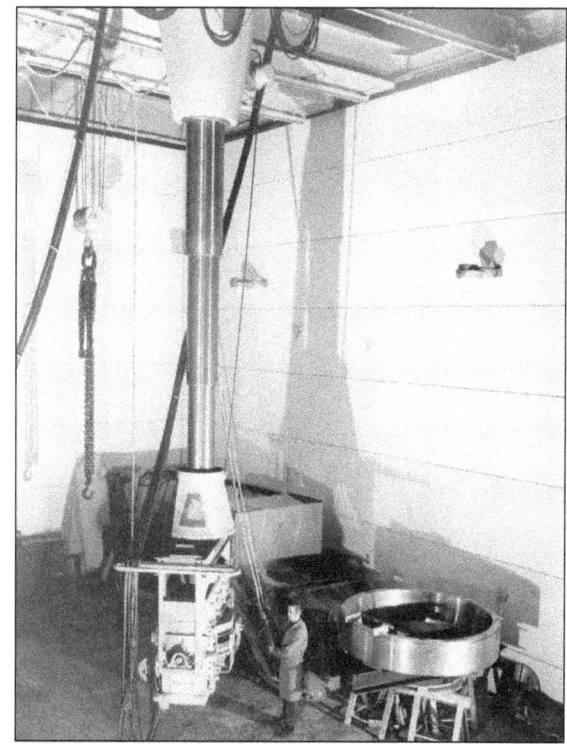

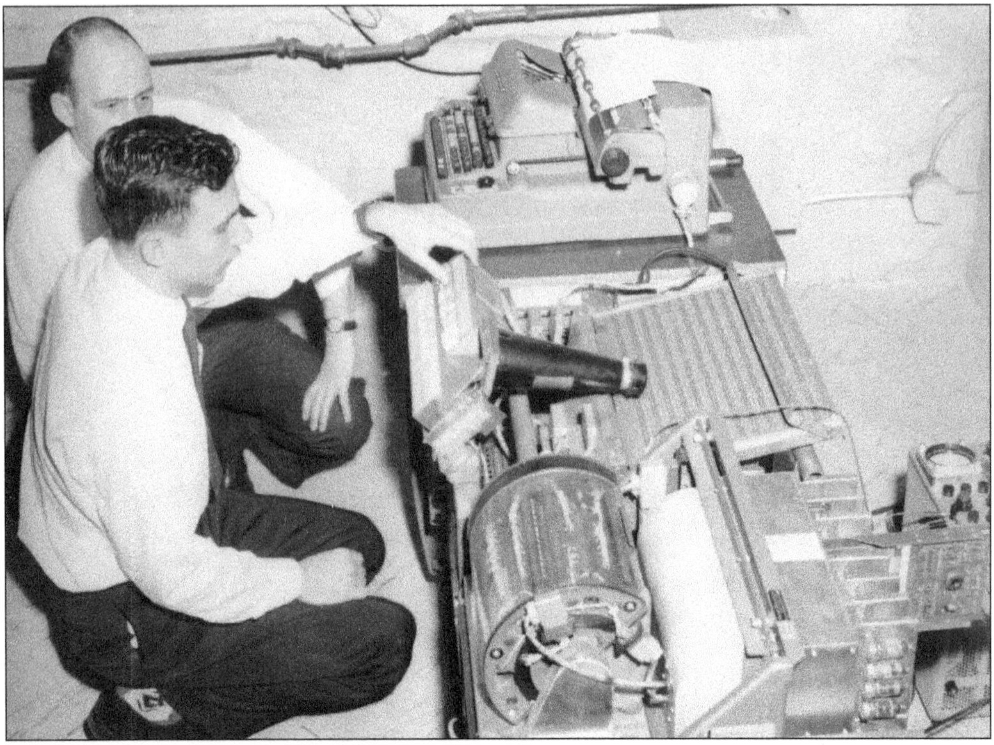

Oscar Werwath had founded the school on the emerging field of electricity. Some 50 years later, Karl Werwath would have the unique opportunity to bring the Milwaukee School of Engineering into the computer age. One of the school's first computer systems, a Royal McBee LGP-30, can be seen in these two images from 1959. It boasted a 4,096-word magnetic drum memory and used 113 electronic tubes and 1,450 diodes. The Royal McBee weighed 800 pounds and was one of the very first off-the-shelf computers commercially available. The purchase of this machine kicked off the Computer Technology Curriculum at the Milwaukee School of Engineering in 1958.

The purchase of a Burroughs B5500 computer and this IBM 1620 Data Processing System in the late 1960s led to the expansion of the Milwaukee School of Engineering Computation Center in 1967. This computation center allowed Milwaukee School of Engineering students to train in many phases of computer technology, operation and servicing, research, design, and development as well as installation and production. Karl Werwath knew that computers were helping to shape the future and he provided his students with a wide array of computer equipment for their study of logic circuits, data flow, logic systems design, mathematical analysis, and programming. The center's Burroughs B5500 computer was operated by Time Sharing Systems Inc. and was to be part of the most advanced and largest remotely accessible time-sharing system for information processing and problem-solving in Wisconsin. In addition to student use, the Computation Center was also made available to the business and industrial community for sponsored research projects.

The Werwath siblings (from left to right) Karl, Hannah Swart, Greta Murphy, and Heinz hold a plaque designating the German-English Academy building an official landmark of the city of Milwaukee in 1971. The building would change hands a few times after this, but this important part of Milwaukee School of Engineering history was returned to the school by Eckhart Grohmann in 2003. Karl Werwath's presidency at the school was marked by grand visions and campus expansion. His unwavering focus on aligning the school with industry partners helped to position it as a place of technological education. The opening of several specialized institutes during his presidency shifted the school from its core of practical education to that of a more research-focused institution. Karl Werwath retired from the presidency in 1977, only two years before his death in 1979.

*Eight*

# CURRENT EVENTS
## MSOE TODAY

The Milwaukee School of Engineering was born out of the innovative spirit of Oscar Werwath and his dedication to providing students with a unique and practical education. His son Karl continued the traditions his father set forth and brought the campus into the modern era. The school was built on a foundation of practical education, technological marvels, individual student attention, partnerships with industry, and a firm dedication to sharing its collective knowledge with the public. These values have kept the original pioneering spirit of the school alive, even as it has grown into an acclaimed university.

Class offerings have shifted and changed with the times, but school leaders work hard to ensure each one aligns with the skills required for success in the workplace. Engineering remains a strong focus and keeps with the tradition of practical applications and experience-based learning. Modern laboratories and equipment are mainstays of the curriculum, just as they were in the early days of the school. Course offerings have been expanded in recent years, with added majors in business and nursing, which have thrived in the technology-focused environment.

The values and interests Oscar Werwath held dear when he founded the school can still be seen today. The school has an individual-focused community feel, with small class sizes and personalized support options, which mirror the early days of Oscar Werwath getting to know each of the students by name. Radio remains deeply rooted in the school's history. Early clubs, hobbyists, and stations paved the way for the current campus radio station. WMSE has entertained listeners with a unique blend of programming for over 35 years. Even the senior design show is reminiscent of the first open houses hosted at the Winnebago Street building where students were encouraged to show the products of their coursework to the interested public.

By adhering to the values that made Oscar Werwath's vision so unique over a century ago, the Milwaukee School of Engineering stays aligned with the next big innovation.

The mace pictured here was constructed in 1965 and gifted to the Milwaukee School of Engineering by that year's graduating class. It serves as a symbol of authority at the school and plays a role in commencement and other official ceremonies. The use of this particular mace was short-lived, as a new version, designed by Prof. Paul Feuerstein, has been used in recent inaugurations and commencement ceremonies.

Dr. Robert Spitzer became the third president of the Milwaukee School of Engineering after the retirement of Karl Werwath. Dr. Spitzer's presidency was marked by growing enrollment numbers and expansion of campus facilities. Pictured here in 1977, Dr. Spitzer accepts the mace to mark the transition of presidential power.

In 1991, Dr. Hermann Viets became the Milwaukee School of Engineering's fourth president. Dr. Viets brought a background in aeronautics engineering and academics. Under his leadership, the school experienced expansions of both its physical campus and available degree programs. Dr. Viets served as president of the school for 24 years and retired in 2015. Here, he receives the mace from previous president Dr. Robert Spitzer.

Dr. John Walz became the Milwaukee School of Engineering's fifth president in 2016. In this image, previous president Dr. Hermann Viets passes the symbolic mace to him. President Walz has a background in chemical engineering and previously held leadership roles at the University of Kentucky, Virginia Tech, and Yale.

In recent years, the Milwaukee School of Engineering has focused on educating determined, driven students in the engineering, business, and nursing fields. The school offers both bachelor's and master's degree programs in each of these areas. Oscar Werwath's original emphasis on real-world training and practical education remains a part of the academic curriculum today, providing students a unique blend of theory and practice. Students focus on their academic pursuits in small classrooms and laboratory settings, and the school makes an effort to provide state-of-the-art equipment on which they can practice. Industry partnerships also continue to be a focus of the school, which allow students to interact with local business leaders to make important connections. Pictured is a 2007 promotional flier for the school that highlights its dedication to collaboration, engagement, innovation, and connections with the community.

Classes in supervisory management began at the Milwaukee School of Engineering in the 1960s. This course of study would eventually lead to the establishment of the Rader School of Business in 1999, which offers degrees in business and management for both local and global business markets. This is a view of Rosenberg Hall, home of the Rader School of Business.

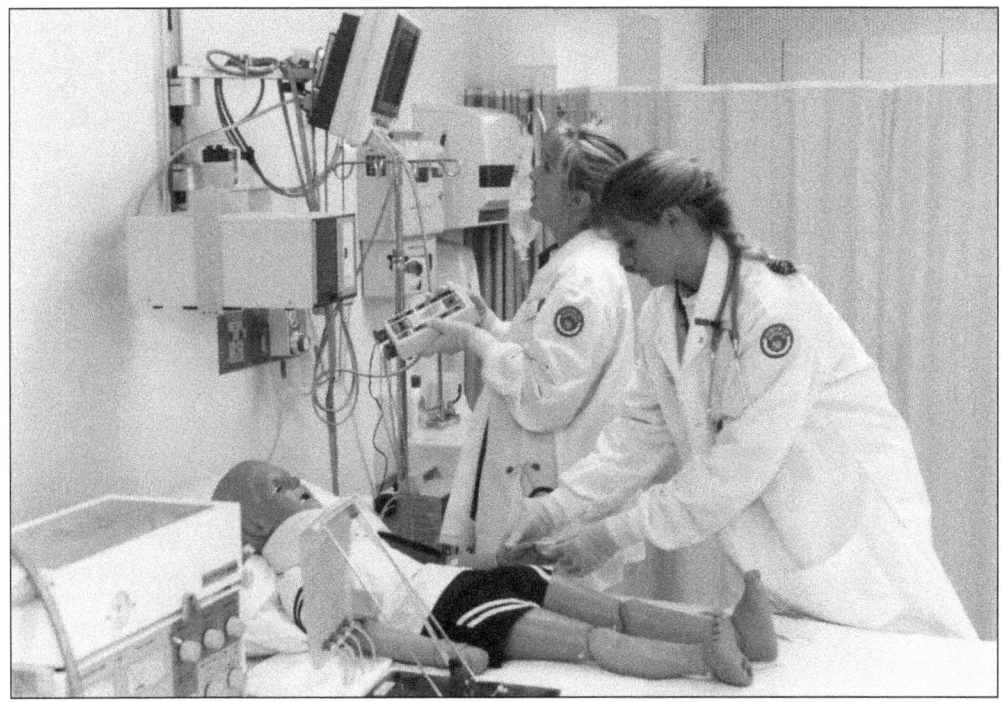

The School of Nursing was established at the Milwaukee School of Engineering in 1995 following an agreement with the Milwaukee County General Hospital School of Nursing. The decision was made to continue the tradition of the original nursing school dating back to 1888 while opening up a new approach to nursing education at a technologically focused school. Pictured is the first nursing laboratory built following the acquisition of the nursing school.

Similar in nature to Oscar Werwath's early exhibitions in the Winnebago Street building, the Senior Design Show is an annual event at the Milwaukee School of Engineering. Early during the previous academic year, students are encouraged to choose a topic in their academic field to research and explore with a group of peers. During this event, students present the results of their year-long projects to members of the public.

Pictured here is the current home of Milwaukee School of Engineering radio station WMSE. The station is a modern continuation of Oscar Werwath's early experiments with wireless technology at the school. Students with an interest in radio technology and broadcasting have always been able to find like-minded colleagues on campus, and WMSE has flowered out of several different groups and clubs. The station celebrated its 35th anniversary in 2016.

In 2003, the Milwaukee School of Engineering celebrated its 100-year anniversary. Many events were held throughout the year to mark the centennial. This image shows two students gifting a time capsule to school president Hermann Viets. This time capsule contains items collected by students throughout the year and is set to be opened at the school's 150-year anniversary in 2053.

The Milwaukee School of Engineering has had its share of challenges during its more than a century of educating young minds. While the technology has changed over time and leaders have come and gone, it is still steadfastly dedicated to the spirit of innovation that Oscar Werwath first championed so many years ago.

Visit us at
arcadiapublishing.com

www.ingramcontent.com/pod-product-compliance
Lightning Source LLC
Chambersburg PA
CBHW060938170426
43194CB00027B/2986